Gas Well Testing Handbook

Gas Well Testing Handbook

Editor

Mamta Sahu

Gas Well Testing Handbook
Edited by **Mamta Sahu**

Printed in 2017

ISBN: 978-1-68117-386-3

Library of Congress Control Number: 2015941574

© 2016 by
SCITUS Academics LLC,
616, Corporate Way, Suite 2, 4766,
Valley Cottage, NY 10989

www.scitusacademics.com

Contents

Preface vii

Chapter 1 A Survey of Methods for Gas-Lift Optimization 1
           Kashif Rashid, William Bailey, and Benoît Couët

Chapter 2 Gas Production from a Cold, Stratigraphically-Bounded
           Gas Hydrate Deposit at the Mount Elbert Gas Hydrate
           Stratigraphic Test Well, Alaska North Slope:
           Implications of Uncertainties 45
           G.J. Moridis, S. Silpngarmlert, M.T. Reagan,
           T. Collett, and K. Zhang

Chapter 3 Simulation of Gas Transport in Tight/Shale Gas Reservoirs by a
           Multicomponent Model Based on PEBI Grid 99
           Longjun Zhang, Daolun Li, Lei Wang, and Detang Lu

Chapter 4 Formation Pressure Testing at the Mount Elbert Gas Hydrate
           Stratigraphic Test Well, Alaska North Slope: Operational
           Summary, History Matching, and Interpretations 125
           Brian Anderson, Steve Hancock, Scott Wilson, Christopher Enger,
           Timothy Collett, Ray Boswell, and Robert Hunter

Chapter 5 Latest Development on Membrane Fabrication for
           Natural Gas Purification: A Review 163
           Dzeti Farhah Mohshim, Hilmi bin Mukhtar, Zakaria Man,
           and Rizwan Nasir

Chapter 6 Parametric Investigation of Well Testing Analysis in Low
           Permeability Gas Condensate Reservoirs 185
           Hossein Mohammadi, Mohammad Hossein
           Sedaghat, and Abbas Khaksar Manshad

Chapter 7   **Analyzing Axial Stress and Deformation of Tubular for Steam Injection Process in Deviated Wells Based on the Varied (T,P) Fields**... 217

Yunqiang Liu, Jiuping Xu, Shize Wang, and Bin Qi

Citations.. 241

Index... 245

Preface

Gas well testing, pressure transient analysis techniques, and analytical methods required to interpret well behavior in a given reservoir and evaluate reservoir quality, simulation efforts, and forecast producing capacity. A highly practical edition, this book is written for graduate students, reservoir/simulation engineers, technologists, geologists, geophysicists, and technical managers. The author draws from his extensive experience in reservoir/simulation, well testing, PVT analysis basics, and production operations from around the world and provides the reader with a thorough understanding of gas well test analysis basics. The main emphasis is on practical field application, where over 100 field examples are resented to illustrate basic methods for analysis. Simple solutions to the diffusivity equation are discussed and their physical meanings examined. Each chapter focuses in how to use the information gained in well testing to make engineering and economic decisions, and an overview of the current research models and their equations are discussed in relation to gas wells, homogenous, heterogeneous, naturally and hydraulically fractured reservoirs.

Editor

Chapter 1

A Survey of Methods for Gas-Lift Optimization

Kashif Rashid, William Bailey, and Benoît Couët

Uncertainty, Risk & Optimization, Schlumberger-Doll Research, Cambridge, MA 02139, USA

ABSTRACT

This paper presents a survey of methods and techniques developed for the solution of the continuous gas-lift optimization problem over the last two decades. These range from isolated single-well analysis all the way to real-time multivariate optimization schemes encompassing all wells in a field. While some methods are clearly limited due to their neglect of treating the effects of inter-dependent wells with common flow lines, other methods are limited due to the efficacy and quality of the solution obtained when dealing with large-scale networks comprising hundreds of difficult to produce wells. The aim of this paper is to provide an insight into the approaches developed and to highlight the challenges that remain.

INTRODUCTION

The introduction of lift gas to a non-producing or low-producing well is a common method of artificial lift. Natural gas is injected at high pressure from the casing into the well-bore and mixes with the produced fluids from the reservoir (see Figure 1). The continuous aeration process lowers the effective density and therefore the hydrostatic pressure of the fluid column, leading to a lower flowing bottom-hole pressure (P_{bh}). The increased pressure differential induced across the sand face from the in situ reservoir pressure (P_r), given by ($P_r - P_{bh}$), assists in flowing the produced fluid to the surface. The method is easy to install, economically viable, robust, and effective over a large range of conditions, but does assume a steady supply of lift gas [1]. At a certain point, however, the benefit of increased production due to decreased static head pressure is overcome by the increase in frictional pressure loss from the large gas quantity present. This has the effect of increasing the bottom-hole pressure and lowering fluid production. Hence, each well has an optimal desirable gas-lift injection rate (GLIR). However, when the entire gathering network is considered, the optimal gas-lift injection rate differs from that which maximizes individual well production due to the back pressure effects (the pressure drop observed across flow lines due to common tie backs further downstream) imposed by connected wells further downstream.

As a field matures, the greater demand for lift gas in conjunction with limitations imposed by existing facilities and prevailing operating conditions (compression capacity, lift gas availability, well shut-in for workover, etc.) can prevent optimal production from being achieved.

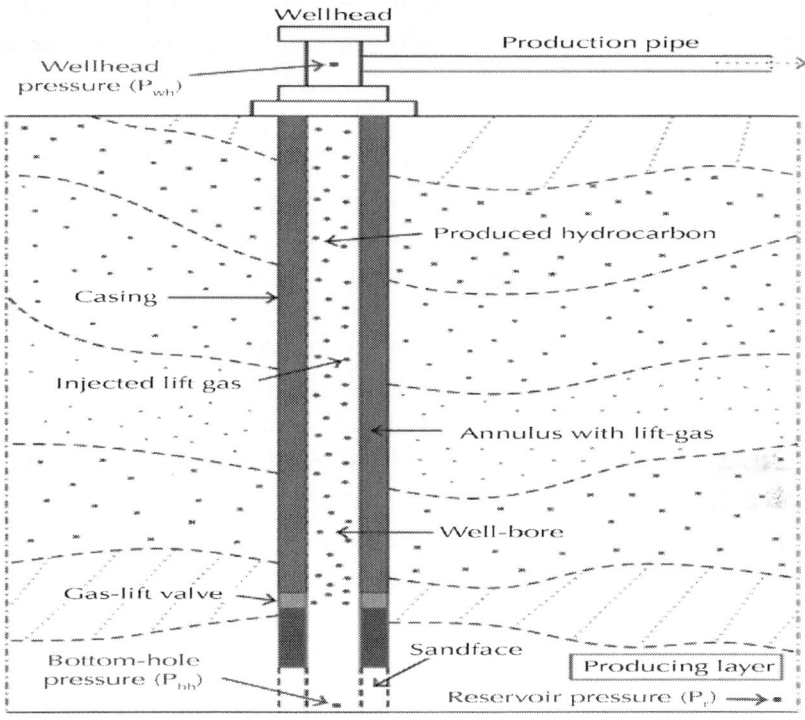

Figure 1: Gas Lift Well Schematic.

In the absence of all operating constraints, other than the available lift gas, it is necessary to optimally allocate the available lift gas amongst the gas-lifted wells so as to maximize the oil production. This is the most basic definition of the gas-lift optimization problem and is equivalent to an optimal allocation problem. Consideration of additional operating constraints, choke control for well-rate management and the treatment of difficult to produce wells, gives rise to a broader problem definition. In general, either definition can additionally accommodate an economic objective function, by inclusion of production or injection cost factors. Although the choice of objective function has been stated as the differentiator between the methods developed by some [2], in actuality, most methods can handle either definition and should not be categorized on this basis.

It is worth noting that generally the gas-lift design problem, that is, deciding the gas-lift valve (GLV) number and their depths, is excluded from the allocation problem, largely because the well configuration is already complete by the production stage and is considered invariable. Additionally, lift gas injection normally takes place at the deepest valve at the available injection pressure, where the depths have been set in advance. This does not preclude well redesign scenarios (workovers) that could yield better production results, whose optimization may improve. For example, Mantecon [3] focused on studies to increase oil production by redesigning individual wells using data gathered at the well site. Individual well performance improvement was considered a necessary requirement prior to field-wide optimization. This involved reevaluating depth of injection, GLIR, injection pressure, GLV size, packer installation and even changing the gas-lift type to intermittent (IGL) or plunger gas-lift (PL). The installation of packers enabled injection pressure increase and deeper GLV activation, collectively leading to increased production within facility handling limits. Although such improvements increase well stability and allow wells to produce more readily, the need for optimal allocation is not obviated. In fact, it is for the reason that the handling facilities and operating resources are constrained that an optimal allocation is required and is most effective when well operation is stable.

In the following sections, a review of methods and techniques developed for the gas-lift allocation problem is presented. These concern, in chronological order of presentation, the generation of gas-lift performance curves from well test data, single-well nodal analysis and sensitivity studies, pseudo steady-state curve-based models that neglect well interactions, steady-state solutions based on network simulators, coupled reservoir and surface facility simulation, ultimately, leading to a fully integrated asset modeling approach. For the benefit of the reader, the key developments are summarized in Table 1 together with the main advantages and disadvantages which will be discussed next.

Table 1: Key developments. The evolution of approaches developed for the treatment of the gas-lift optimization problem

Merits	Limitations
Performance curve generation	
Provides well production	Well test requirements
relationship with GLIR	Well test data quality
Nodal analysis	
Well model simulation	Fluid data assumptions
Multi-phase flow modeling	P and T assumptions
Performance curve generation	Primarily for single well
Curve-based models	
Fast, analytical models	Neglect well interations
Considers all wells	Curve fitting and quality
Simple to evaluate	Pseudo steady state solution
Network simulation	
Rigorous simulation models	Evaluation cost
Includes well interactions	Model smoothness
Handles looped models	Steady state solution
Handles facility components	Gradient information
Coupled simulation	
Detailed coupled system	Coupling scheme
Rigorous interaction	Model robustness
Includes transient effects	High computation cost
	Gradient information
Integrated asset modeling	
Comprehensive system	Coupling procedures
dynamics and interactions	Model robustness
Simulation over asset life	High computation cost
	Gradient information

WELL GAS-LIFT PERFORMANCE

Actual well investigation concerns physical well tests conducted at the well site. Fluid composition, PVT and other related tests provide information about the conditions of the well and its potential productivity. In addition, step-rate gas injection tests can provide an accurate description of the behavior of fluid production with increasing lift gas injection.

The nature of single-well testing naturally gave rise to the development of tools to model the behavior of a single well, given certain input parameters to define the completion, fluid composition, pressures, and temperatures, both at the wellhead and at the point of contact with the reservoir. These Nodal Analysis tools enable a computational model of the well to be defined, with either a simple black oil or a more detailed compositional fluid description, commencing from the sand face, through the perforations, from bottom hole, through tubing, to the wellhead and further downstream to a delivery sink [4]. This model can then be used to predict the behavior of multi-phase flow through the well (see Figure 1) and the more representative it is of the prevailing conditions the better. Thus, to overcome the complication of performing costly and time-consuming step-rate tests in a field with many wells, these tools can be used to provide lift performance curves for all lifted wells [5, 6]. Clearly, the analysis of a single well provides an incomplete picture with respect to the entire field performance and especially for optimal gas-lift allocation. However, it serves two important purposes. Firstly, all network simulators, which couple the behavior of individual wells in a field by a common gathering network, are based on such single-well models. That is, in solving a network solution, the underlying multi-phase fluid flow behavior is calculated using nodal analysis tools. Secondly, in the absence of actual step-rate test data, lift performance curves are estimated by running sensitivity models over each of the individual wells. It is these lift performance curves that are used in gas-lift optimization studies. This is undertaken correctly, when the effect of interdependent wells is accounted for by updated lift performance curves at new conditions, but incorrectly when the resulting pseudo steady-state solution is accepted after only one cycle. The latter is a common assumption in many papers, often for the sake of simplicity, if not by oversight. The complete, or final, steady-state solution is one derived from the network simulator in which a rigorous pressure balance is achieved over all nodes in the network after the allocation of lift gas has been made to the wells. This is necessary since the back pressure imposed by the injection of lift gas in one well will affect the production from all connected wells. Hence, for optimal production the lift gas must be suitably allocated whilst accounting for well interaction. Lastly, facility handling constraints (i.e., constraints imposed at a field-wide level, e.g., maximum water production, etc.) are also only applicable if the entire network model is considered.

SINGLE WELL ANALYSIS

The use of nodal analysis to generate the gas lift performance curve of a single well based on actual pressure and temperature surveys along with a suitable multiphase flow correlation is well established [7–9]. The optimum GLIR is often simply set to that furnishing the highest production rate on the gas-lift performance curve (GLPC) (see Figure 2). The maximum GLIR and maximum oil rate can be used to establish the optimal valve depth setting and the wellhead pressure (P_{wh}) [9]. However, the single-well configuration, considered in isolation of other wells, is not a field gas-lift optimization solution.

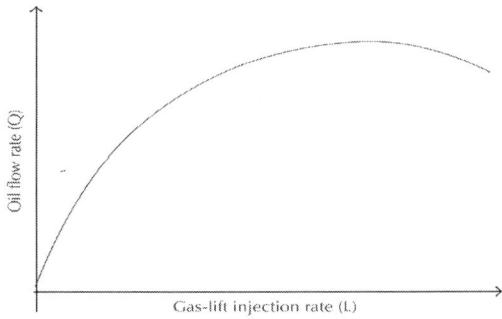

Figure 2: Gas Lift Performance Curve at a given Wellhead Pressure (P_{wh}).

A more accurate well model, based on mass, energy, and momentum balance, was proposed by Vazquez-Roman and Palafox-Hernandez [10]. A single-well case was examined to determine the injection depth, pressure, and the amount of gas injection, using a commercial optimization program based on a hybrid interior algorithm. The results were reported as being more accurate and therefore better suited for field-wide simulation studies, than the standard nodal approach. However, no results were reported for a field-wide application. Note that the use of compositional models over simple black oil models is also recommended practice for better accuracy [11, 12]. Indeed, the best possible well model, yielding representative gas-lift performance curves, is desirable for both simulation and optimization purposes.

Dutta-Roy and Kattapuram [13] investigated the introduction of lift gas to a single well using nodal analysis. The back pressure effects imposed by gas injection on two wells were considered, prior to an investigation of a 13 well network. It was noted that single-well methods were inadequate for analyzing a production network of many gas-lifted wells and that a general network solver is thus required. That is, optimizing wells systematically, but on a well-by-well basis, will not guarantee that an optimal solution will be obtained for the entire network. Equally, Mantecon [3] noted that for field-wide optimization, accurate gas injection and liquid production estimates are desired, but more importantly, as the conditions in one well affect other connected wells, a computer simulation is required to effectively account for the interactions. This cannot be achieved manually.

Bergeron et al. [14] used step-rate well test data to obtain the characteristic lift performance curve for a single offshore well. A remotely actuated controller was used to manage the gas-lift injection rate in real time for production maximization, cost reduction and mitigation of blockages resulting from flow line freezing. Accurate flow rate and injection measurements were deemed necessary to ensure accurate model interpretation and optimization, while also ensuring stability from heading and slugging effects. The single-well optimization scheme adopted set the unconstrained optimum with an unlimited supply of lift gas or established the injection rate for production maximization in the presence of operating constraints, including the amount of lift gas available. Hydrate formation and low ambient temperatures, together with changes in lift gas supply, pressure, or quantity can lead to suboptimal levels of gas injection and consequently, a drop in cumulative well productivity. Continuous monitoring and sustained optimization were considered imperative to ensure that the well operated at maximum efficiency the majority of the time leading to an observable increase in total production. It was noted that to connect-up multiple wells, a more sophisticated optimization routine was necessary to handle the inter-related well allocation problem. Although only a single well was considered and the optimization scheme was limited to setting the lift gas rate resulting in the highest flow rate, the elements necessary for remote real-time field-wide gas-lift optimization are evident.

PSEUDO STEADY-STATE MODELS

In the previous discussion, it should be noted that direct step-rate well-site data provides a field model of sorts (case 1). Using PVT and composition data to generate gas-lift performance curves using nodal analysis tools can provide independent single-well models (case 2). Thirdly, employing a rigorous multi-phase network simulator, a composite full-field model is obtained (case 3). Whereas the latter implicitly models the back pressure effects imposed, leading to a steady-state solution, the first two approaches choose to neglect them. Hence, such methods (case 1 and case 2), though perhaps extendible, provide only a pseudo steady-statesolution.

In the following, methods established for case 1 & case 2 are considered as one, in that the descriptive well performance curves employed for any well may be obtained from a single-well model or exemplifying actual well behavior, directly from well-site step-rate tests. The clear limitation of case 1 and case 2 is that they do not account for the true back-pressure effects imposed by gas injection in one well on the other wells connected in the network.

Although such methods result in pseudo steady-state solutions, unless explicitly stated, this limitation can be overcome by iterating on the procedure presented with updated well data and gas-lift performance curves. In general, more detailed simulation models, including the reservoir and the process plant, can be defined with increasing complexity as case 4 and case 5, respectively. Note that the latter embodies the entire asset.

Simmons [15, 16], and Kanu et al. [17] describe the generation of well lift performance curves. Typically, tubing intake pressure as a function of the production rate is plotted for varying gas liquid ratios (GLR). The well inflow performance curve (IPR) is superimposed and the intersecting points with the tubing intake pressure curves identify the highest production possible for any given GLR. This data gives rise to the characteristic gas-lift performance curve (GLPC) for each well [18, 19] (see Figure 2). The most efficient injection rate was noted as the point at which the incremental revenue is equal to the incremental cost of injection and not simply when production is maximized [15–18, 20].

Simmons [15, 16], Redden et al. [20], and Kanu et al. [17] accommodate economic factors by including the net gain (or profit) from oil and gas production, and the costs associated with gas compression, gas injection and water disposal, and so forth. For incremental profit as a function of gas injection Simmons [15, 16] considered the difference between the revenue and cost. The solution was defined at the point where the incremental profit is zero (see Figure 3(d)) and was referred to as the maximum daily operating cash income (OCI). Redden et al. [20] and Kanu et al. [17] similarly converted the GLPC into monetary units. Redden et al. [20] proposed the optimum point to occur when the slope of the revenue versus cost curve is one (see Figure 3(b)). Kanu et al. [17] referred to this solution on the GLPC as the economic slope (see Figure 3(a)) and the equilibrium between revenue and cost as the economic point. Under consistent revenue and cost factors, each of the aforementioned schemes gives rise to the same solution, including profit maximization as a function of gas injection at zero slope (see Figure 3(c)).

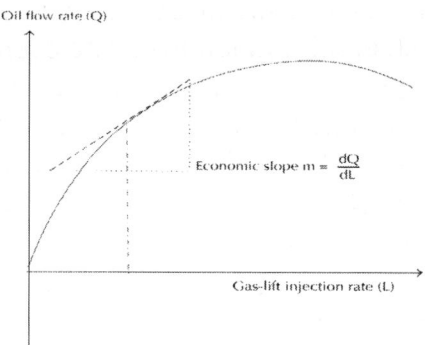

(a)

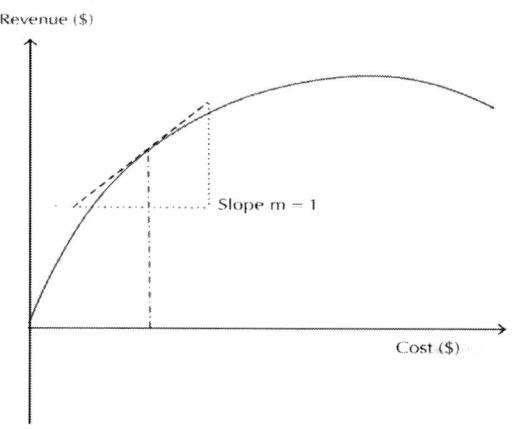

(b)

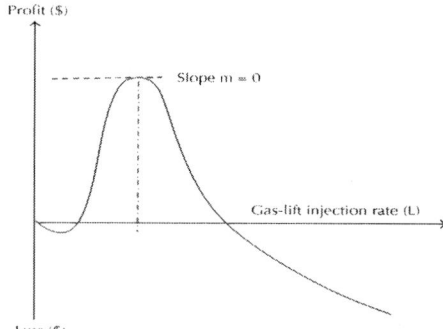

(c)

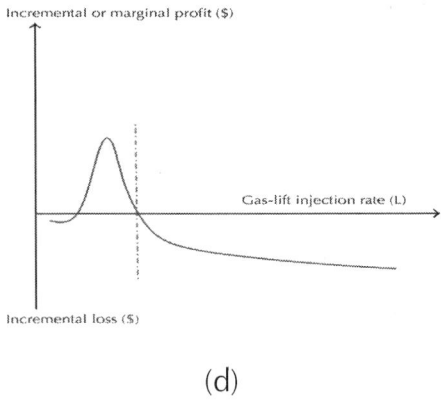

(d)

Figure 3: Performance curve representation and solution schemes.

Kanu et al. [17] are often cited for the "concept" of the equal-slope solution. Note that the term "method" is not used here, as the method implemented to obtain that solution can vary. Moreover, the equal-slope concept arises only if the average economic slope is used based on averaged well properties (e.g., an average water fraction over all wells). In general, given the differing conditions of each well, each economic slope will differ. Kanu et al. [17] noted that, while the use of average properties simplifies the procedure, individual economic slopes should be used for increased accuracy. In actuality, the equal-slope concept features more palpably in the work of Redden et al. [20] (solving for unitary slope on the revenue versus cost curve; see Figure 3(b)) and even earlier in the work by Simmons [15, 16] (solving for zero incremental profit, see Figure 3(d)). Both employ the equal-slope concept, but as observed by Clegg [21], were not cited in the work of Kanu et al. [17,22].

The reciprocal of the slope of the GLPC gives the incremental gas-oil-ratio (IGOR) [23]. This notion was used to define an optimal solution based on equal IGOR for each well, referred to as the marginal GOR (MGOR) by Weiss et al. [24] and is the same as the equal-slope concept. A proof of optimality for a two-variable model based on substitution is provided by Weiss et al. [24] and is stated to be inductively correct for higher dimensions. Notably, the optimal condition is based on satisfying the maximum gas capacity constraint. It overlooks the notion of an over-abundance of available lift gas and therefore the likelihood of poor lift efficiency due to excessive lift gas utilization.

In recent work [25–27], the Newton Reduction Method (NRM) was presented as a means for optimal allocation. The method reduces the original problem to a solution of a composite residual function of one variable by treating the available lift gas as an equality constraint. The method is derived from a strict mathematical formulation and is provably optimal, assuming the GLPC are convex. The method is fast and returns an equal-slope based solution.

Simmons [15, 16] noted that GLIR for maximum daily operating cash income (OCI) (the point at which the incremental profit is zero, see Figure 3(d)) is lower than the GLIR for maximum oil production and the GLIR for maximum present value operating cash income (PVOCI) (when well reserves are also considered) is lower still. The latter (PVOCI) is expressed as the desired quantity, though it was observed that the PVOCI was insensitive to gas-lift rate variation above and below the optimum GLIR. However, exceeding the optimum GLIR causes facility overloading and yields lower PVOCI. The solution method is based on an incremental allocation of small units of gas until either maximum production, OCI or PVOCI are obtained. As the same slope is obtained for each well at the optimal point, this is an equal-slope based approach. It is interesting to note that many early implementations for gas-lift optimization adopted the same concept and a similar heuristic scheme [28, 29]. The gas-lift allocation procedure within the ECLIPSE reservoir simulator also uses a similar approach [30].

The multi-phase flow program and allocation procedure by Simmons [15, 16] proved useful in reducing the gas lift requirement to low producing wells when the gas supply was limited. A 20-well system was considered, however the predicted performance observed differed from reality due to a number of reasons. These included over-injection for well stability, misestimation of injection rates or wellhead pressures, inaccuracies in well data, correlation error, model parameter uncertainty and the heading problems observed. The last item concerns the effects of oscillatory fluctuation of pressure in a well. In general, these issues will confound any simulation-based approach and signifies the need for accurate data with which to develop the underlying models. As previously noted, a stable well configuration is desirable before performing optimization. Hence, overcoming heading for greater well stability may require well redesign. In this regard, Mantecon [3] and Everitt [31] both employed measures to identify and mitigate these issues (using redesign and intervention) prior to successful gas-lift allocation.

The Redden et al. [20] approach involves finding point at which the gradient of the monetary-based GLPC is one. However, no mention was made of the procedure to achieve this. If the lift gas allocation solution exceeds the available gas or separator capacity, a ranking of wells is made and the lift gas is removed from the lowest producing wells. The incremental reduction is continued until separator and compression limits have been met. This is not unlike the allocation scheme employed by Simmons [15, 16] however, in reverse. Such heuristic-based schemes are not proven to be optimal especially in the presence of multiple constraints and specifically in the given case where naturally flowing wells were excluded from receiving any lift gas at all. Results of a 10-well model were presented with and without capacity limitations in place.

Kanu et al. [17] solved for the economic point using a method derived graphically. The optimal operating condition is said to occur when the incremental revenue from production is equal to the incremental cost of injection in each well. The production and gas-lift rates for a range of slope values are estimated for each well. These give rise to slope versus production and slope versus gas-lift rate relationships. The economic point for each well is established and the associated lift-rate and production values are obtained. Total production and the total lift gas used are established by summing the individual well solutions. The same can be performed for all slope values, allowing the relationships of Total Production and Total Lift gas to be plotted with respect to the slope. With a limited supply of gas, the amount of gas available will indicate the expected slope value from the total lift gas versus slope plot. The associated production value can be obtained from the Total production versus slope plot for the given economic slope. Similarly, the individual well responses can be read from the particular well plots.

It is worth noting that the economic slope solution is greater than the zero gradient necessary for maximal production, which indicates the benefit of optimizing for economic performance and not simply production. A 6-well model was presented by Kanu et al. [17] with an unlimited lift gas supply using actual and an average estimate of the well properties. While the latter simplifies the evaluation process, the solution is less accurate. The constrained lift gas solution implicitly returns an average economic slope, leading to an allocation that is not strictly correct or optimal. In general, the procedure cannot easily handle additional constraints and can prove unwieldy for high-dimensional problems and cases where the curves have to be regenerated frequently due to changing well conditions. For this, the authors note that an automated procedure is necessary.

The equal-slope concept has been adopted in several other works [28, 31–33]. Typically, and incorrectly as the back pressure effects are neglected, when there is an unlimited supply of lift gas, the wells are often set to the GLIR that simply maximizes production in each well. With a limited supply, the equal-slope concept is employed.

Edwards et al. [32] fit each GLPC generated from a multi-phase flow simulator with a polynomial and identified the zero-gradient point of each well as the unconstrained optimum. In the limited case, an equal-slope solution was obtained using an iterative scheme, of which no particular implementation details were provided. Results for an 8-well model were presented.

Ferrer and Maggiolo [33] discussed the use of two computer models. The first, to interpret well behavior, performs diagnostics and well redesign. The second, to generate well relationships (GLPC) and perform optimization. The equal-slope concept is employed if the gas supply is limited, however, only a single-well model was evaluated. Everitt [31] used actual data to define the lift performance curve for each well. Although this can be costly and time-consuming, in comparison to simulation-based relationships, the curves obtained were representative of actual field conditions observed. The work focused on individual well performance identification and presented measures to alleviate the heading problems observed in several wells. That is, the oscillatory well fluctuations that occur due to changing pressure conditions in a well. The decision to close high water-cut wells helped reduce compression demand and cost.

Schmidt et al. [28] also employed the equal-slope concept for allocating lift gas, but details of the method were not provided.

The gas-lift optimization method cited by Chia and Hussain [29] is also based on the equal-slope concept. However, instead of simply using lift curve data, a simplified black oil network tool (GOAL) [34], employing a family of lift performance curves to describe the behavior of each of the wells, was used. A fast flow rate and pressure balance calculation enabled the back-pressure effects between connected wells to be accounted for during the allocation process. The available lift gas is discretized and allocated incrementally to the wells with favorable production gradients until an equal-slope solution is obtained. The method is fast, but has the disadvantage of simplifying the fluid compositions of the wells, leading to some loss of accuracy in the solution.

The allocation of lift gas to a number of central processing facilities was considered by Stoisits et al. [23]. The composite lift performance curve was presented as the sum of the individual GLPCs for each of the wells connected to them. The MGOR concept, equivalent to the equal-slope solution, was used to allocate the available lift gas, but implementation details were not provided. A neural-network model was employed to simulate the behavior of the surface line hydraulics.

Weiss et al. [24] also employed the MGOR concept, in conjunction with developing a dedicated gas-lift allocation correlation based on empirical data. This provides the optimal GLR as a function of particular well properties, but treats only a single well in isolation. Stoisits et al. [35] compared the empirically derived correlation based on a log equation from Weiss et al. [24] with the adaptive non-linear neural network model from Stoisits et al. [23]. The reported results differed markedly, possibly due to the difference in solution from the nodal analysis based runs used to train the neural network and the empirical data used to build the correlation. The latter is also potentially less versatile over a large range of conditions, as it conditioned to the well data employed.

Nishikiori et al. [2] presented the application of more standard non-linear optimization procedures for the gas-lift optimization problem. A first-order quasi-Newton method was developed. This is known to present super-linear convergence at a starting point close to the optimal solution. For this reason, three methods for automatically providing more informed GLIR starting points were considered. These include uniform lift gas distribution and the ratio of either the well productivity index (PI) or the maximum liquid production, to the corresponding sum for all lifted wells. The requirement of accurate first derivatives necessitated a costly central-difference scheme. A 13-well system was tested with both limited and unlimited gas availability. The method had the advantage of being fast and efficient.

Mayo et al. [36] used the Lagrange multiplier approach with each of the gas-lift performance curves modeled using second-order polynomials. The formulation gave rise to a convex constraint set and the imposition of Karush-Kuhn-Tucker (KKT) conditions for optimality guaranteed the resulting solution as being globally optimal [37, 38]. The method of solution was fast, though perhaps with some loss of accuracy due to the fitted polynomial, and was used to remotely actuate all 34 gas-lifted well valves through a controller and not just for a single-well as performed by Bergeron et al. [14]. In the unlimited case, the apex point of each GLPC, identifying maximal production, was set and for the limited case, the lift gas deficit (from the optimal lift gas requirement) was shared equally among the wells.

Lo [39] also used a Lagrangian formulation, but with the intent of dealing with multiple constraints. A general non-linear optimizer was employed (of which no details were given) to solve the constrained allocation problem. With a lift gas inequality constraint only, the equal MGOR solution was obtained for production maximization. In the presence of additional production constraints, the MGOR was noted to differ for each of the 20 wells in the model tested. Lo observed, with some foresight, that if, as proposed, the lift performance curves are assumed concave, then the objective function is concave and the feasible region defined by the constraints is convex. Thus, "one possibility for future work… is to devise a faster algorithm to solve the well management problem compared to [the] general non-linear program" employed. In addition, as the back-pressure effects were not considered, the problem should be "solved iteratively until the predicted WHP are consistent." Both these features happen to be the basis behind the gas-lift optimization scheme encompassing the Newton Reduction Method (NRM) [27]. However, while the general non-linear scheme adopted by Lo [39] enables the application of many constraints, including those at the manifold level, the NRM approach is somewhat limited [25].

Fang and Lo [40] and Handley-Schachler et al. [41] both proposed the use of Sequential Linear Programming (SLP) techniques. The lift performance curves were assumed to be piece-wise linear and the constraints were linearized using a first-order Taylor series expansion. The approach can be fast, as it requires only first derivatives, and is able to handle bigger models with relative ease. However, the limitation is that the linear model maybe a poor representation of the highly nonlinear system and non-instantaneously flowing (NIF) wells can be problematic (see Figure 4). Both papers utilize separable programming, the adjacency condition and special order sets in order to qualify the lift performance curve of each well and to optimize the performance of the linear solver [40, 41].

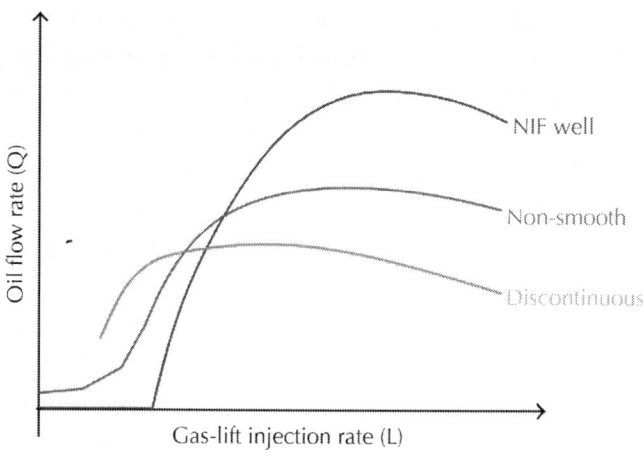

Figure 4: Gas Lift Performance Curves at a given Wellhead Pressure (P_{wh}).

Handley-Schachler et al. [41] presented a 21-well model for a de-bottlenecking study and a 250-well model for produced gas usage optimization problem, including the amount used as injection gas. The objective was to maximize revenue by lift gas allocation, while controlling other factors such as, compressor power, suction and discharge pressures. Fang and Lo [40] examined two full-field test cases, taking account of changing reservoir conditions, to maximize the cumulative oil production.

An improved procedure to generate the GLPC to more accurately match the test data collected was reported by Salazar-Mendoza [42]. On the other hand, Alarcón et al. [19] employed a modified second-order polynomial to fit the well GLPCs obtained from nodal analysis. The addition of a correction term gave rise to a better fit than the use of standard second-order polynomials reported by Nishikiori et al. [2] and Mayo et al. [36]. The standard non-linear problem was posed and the convexity of the curves noted to yield a globally optimal solution. The Sequential Quadratic Programming (SQP) method was used for the solution of the problem, using an approximate Hessian update procedure.

The SQP method is the workhorse of constrained non-linear optimization. A quadratic approximation is made at a point on the Lagrangian function and the constraints are linearized. The resulting quadratic subproblem is solved for the direction of search, along which a line search is performed to minimize the merit function [37,38]. The matrix of second-order derivatives, the Hessian, must be updated to improve the quadratic approximation of the Lagrangian function using a suitable update procedure, for example, the BFGS method [38]. The cycle is repeated iteratively until convergence to a local minimum. The method suffers from the requirement of second-order derivatives, which can be costly and difficult to evaluate. The method can handle constraints well and from a good starting point can exhibit quadratic convergence. However, using approximate Hessian update procedures, only super-linear convergence can be guaranteed [37].

The need for a good GLIR starting condition was addressed by Alarcón et al. with a distribution of available lift gas proportional to the maximum flow rate over the sum of all wells, as also suggested by Nishikiori et al. [2]. For non-instantaneous flow (NIF) wells (i.e., those requiring a minimum level of gas injection before they can flow, see Figure 4), minimum GLIR rates were specified. This has the effect of forcing the wells to produce, but can lead to a suboptimal solution as certain wells unnecessarily consume large amounts of lift gas simply to produce small amounts of oil. Lo [39] noted that NIF wells break the concavity assumption and incorrectly suggested that as long as the minimum GLIR is far from the GLIR required for the unconstrained maximum flowrate, the use of minimum GLIR constraints is reasonable. However the excessive allocation of lift gas to low producing wells that results, prevents better solutions from being obtained.

Alarcón et al. examined test cases of 5, 6, and 13-wells, of which the first two were also considered by Buitrago et al. [43] using a global derivative-free method. In addition, the 6-well problem was addressed by Ray and Sarker [44] using a multiobjective evolutionary approach. However, [19] only compared results from Buitrago et al. [43], from which the test cases were taken. The solutions from Alarcón et al. differed to those from Buitrago et al. [43] somewhat due to the method of solution employed and the curve interpolation scheme used. In general, the improved polynomial fit and the robust SQP solver yielded better results for Alarcón et al. Note, however, that the single NIF well present in the 6-well case was treated in a binary manner, that is, either on or off. The necessary treatment of a problem with a greater number of NIF wells was not made clear.

The complications arising from wells with nonsmooth gas-lift performance curves, discontinuities and indeed with non-instantaneous flow can impede the progress of conventional gradient-based optimization methods (see Figure 4). This results in sub-optimal solution due the poor handling of such wells and due to simplification made by curve fitting. For example, the flat section of a NIF well returns a gradient that is of little use in the optimization procedure. Hence, a minimum lift gas injection constraint is often implemented to overcome this issue, but this leads to an inefficient allocation and a sub-optimal solution, especially with increasing dimensionality. Conventional methods also tend to find only local solutions. To overcome these concerns, several researchers resorted to the use of more robust global search algorithms. Buitrago et al. [43] developed a global derivative-free search algorithm for the gas-lift allocation problem using heuristic measures to evaluate the descent direction. The method explores promising areas of the search space using a predefined number of samples using random search and a clustering method. The method is reported as being able to handle any number of wells, including wells with irregular profiles and non-instantaneous flow (NIF) (see Figure 4). The paper presented test cases of 5, 6, and 56 wells. The last two cases comprised 1 and 10 NIF wells, respectively. The method is compared to solutions obtained using an equal-slope approach. It is interesting to note that the solutions of the reported method by Buitrago et al. [43] are with the NIF wells turned off. However, a comparative study has shown that better solutions are possible with certain NIF wells activated [27]. Hence, although the method is robust, it does not appear to show any advantage in being able to accommodate NIF wells at all. Buitrago et al. [43] proposed the metric of gas lift per barrel of oil at the solution point as a means of comparing solutions for different methods. However, while certain methods allocate all the available lift gas and find only local solutions, others will not. As such, the best comparative metric can only be solution quality, based on the overall objective value (e.g., the total oil production) in the absence of the computational cost necessary to obtain it. Note that many papers provide no function call count as a measure of computational cost or the efficiency of the method presented with increasing dimensionality. Lastly, it should be noted that for the third test case in Buitrago et al. [43], the sum of allocated lift gas is mis-stated. This should be 20,479 MSCF/D for the reported method and 24,661 MSCF/D, which exceeds the quantity of available gas, for the equal-slope solution used for comparison. This changes the metric ratios reported from 0.939 and 1.059, to 0.94 and 1.16, respectively.

The use of a genetic algorithm (GA) for the gas-lift allocation problem for production maximization was reported by Martinez et al. [45]. A genetic algorithm is a stochastic population based global search strategy in which an initial set of candidate seeds, often randomly selected, is evolved over a number of generations using the key operations of crossover, reproduction, and mutation. The fittest candidate in the final gene pool is the solution to the optimization problem. The method is derivative-free, robust and can potentially find a global solution. In addition, both integer and continuous variables can be handled with ease. However, the method has the disadvantage of requiring a great number of function evaluations, which can be computationally expensive. Many variations and extensions exist to speed up convergence, maintain population diversity and to avoid binary string encoding. Algorithm parameter tuning may also be necessary for optimal performance. For example, in Martinez et al. [45], a greedy step was included to prioritize wells for preferential gas lift allocation and constant lift curves were used to test models of 10 and 25 wells. In addition, uniform and average crossover, mutation, short- and long-creep steps were considered for algorithm performance improvement. The method reportedly outperformed individual well schemes for production maximization by twenty percent.

Ray and Sarker [44] posed the gas-lift optimization problem as a multi-objective problem (MOP), with the intent to maximize production while minimizing lift gas usage. In the single objective problem of maximizing production alone, the second objective is treated as a constraint. A variant of the NSGA-II algorithm [46] was employed as a means of maintaining population diversity. The acronym refers to the nondominated sorting genetic algorithm, which indicates that the best solutions identifying the Pareto front are retained [46]. Contrary to Martinez et al. [45], where the robustness of the GA was used to help overcome the irregularities observed in the GLPCs, the lift curves were instead assumed to be piece-wise linear. The 6- and 56-well cases from Buitrago et al. [43] were employed for testing. The derivative-free algorithm was robust and able to deliver solutions to the problems, each with 1 and 10 NIF wells, respectively. They reported improvements compared to the results by Buitrago et al. [43]. However, it should be observed that the simulations were run 96 times each and the best result obtained was used for comparative purposes. If the median or the average result is employed instead, the improvement is eroded or overcome completely.

The use of the MOP solution set to derive a production versus total gas injected profile in Ray and Sarker [44] is useful. However, retention of the solution configuration at each point on the profile is necessary, and no mention is made of this. In general, the MOP solution set is expensive and unnecessary given that much effort is required simply to filter out the dominated set of points. For example, the Newton Reduction Method will return the production versus total gas injected profile at selected values of the total available lift gas directly [25]. The benefit of multi-objective optimization could have been explored with more complex and conflicting operating constraints.

The shortcoming of curve-based network models, while better than undertaking single-well analysis, is evident in their neglect of the back pressure effects imposed by the entire network. The injection of lift gas in one will have an effect on all connected wells and this is not treated when the wells are considered separable.

In addition, significant changes in well condition, bottom-hole pressure, productivity, water-cut or well-head pressure, all necessitate the update of the lift performance curves. As gas injection tends to lower BHP and increase WHP due to the fixed delivery point pressure, performance curve update is imperative. Hence, curve-based models give rise to pseudo steady-state solutions only and these will suffer from significant fluctuation in either well, operating or facility conditions, and with increasing model dimensionality. In the following section, network models are utilized to overcome this shortcoming.

NETWORK-BASED SOLUTIONS

The use of a non-linear constrained optimizer was reported as a necessity by Dutta-Roy et al. [47] when considering fields with many interconnected gas-lifted wells and facility components. A rigorous multi-phase flow network solver was used to solve the pressure and flow rates across the network in conjunction with a SQP solver for the constrained allocation problem. Function evaluation count was not reported, but the benefits of a simultaneous solution for more accurate results was shown [13, 47].

Nadar et al. [48] defined a network system to fully account for the production and gas-lift system interaction simultaneously for a complex offshore gas-lifted operation for production gain and cost reduction. A family of lift performance curves for varying WHP was generated for each well. The surface gas-lift components were modeled in the network simulator, including a detailed compressor model. The curves were assumed piece-wise linear and the SLP method employed for optimization. The separator pressure, gas-lift header pressure and the gas-lift injection rates were assigned as control variables. The approach was tested on 2 configurations of a composite model based on 4 fields, comprising 40 production platforms and 200 wells, for revenue maximization under equipment and network constraints. The method is able to handle large complex looped networks and can treat constraints simultaneously. A number of studies were examined, including gas-lift compressor train shutdown, establishing the gas-lift injection pressure and evaluating the gas transfer across fields. The difficulty of defining the gas-lift performance curves at low injection rates was noted, and thus can affect the solution quality. In general, inclusion of the gas compression and gas injection system enhances the overall merit of the optimization solution, in comparison to the production pipeline model alone.

Vazquez et al. [49] proposed an approach combining a genetic algorithm with the Tabu Search method. The latter performs a local search and retains a set of search points marked as in viable, known as the tabu steps. The search progresses in a direction where the objective function improvement is most likely. The method was tested on a 25-well system and a 5% increase in production was reported compared to the original state. The system considered comprised the production wells, the surface facility model and included a number of operating constraints. Although other forms of artificial lift (sucker rod pump and electric submersible pumps) were considered, the scheme can equally be applied to a gas-lift scenario. The main drawback of the approach was the high function evaluation cost resulting from the global stochastic-based scheme.

To ease the computational burden of simulation cost, Stoisits et al. [50] introduced an adaptive non-linear model to replace the actual network production simulator. The production model comprised the individual wells, surface line hydraulics model, and the production facility model. A GA was combined with a Neural Net (NN) based production simulation model for production optimization. The GA was used to return a global solution, while the NN model provided a fast proxy of the actual objective function once successfully trained. The problem investigated concerned the allocation of wells to production facilities and the lift gas to the wells, under multiple production constraints. The GA was used to find the optimal well activation state and the optimal drill site IGOR (for a collection of wells), which returns the GLIR for each individual producing well. Results from the GA were compared with and without parameter tuning and a production increase between 3% to 9% was achieved. In Stoisits et al. [35], the NN model was compared to the empirical correlation by Weiss et al. [24] using the composite IGOR curve definition developed [23].

An iterative offline-online procedure for gas-lift optimization was developed by Rashid [26, 27]. Gas-lift performance curves are extracted and used to solve the gas-lift allocation problem offline. Contrary to the assertion that spline-fitted GLPCs are unsuitable [19], the method uses smoothing and splines to ensure convexity of both the GLPC and the inverse derivative profiles required by the NRM solver (see Figure 5). The offline solution is passed to the network model [5] for updated wellhead pressures (WHP). The procedure repeats until convergence on WHP has been achieved [27]. This approach has the benefit of speed through the use of well curves, while retaining the rigor of the actual network model. Hence, flow interactions are fully accounted for. Note that the NRM solves a residual equation of one variable assuming GLPC convexity and a strict allocation of all available lift gas [27]. However, a GA solver is also embodied to handle nonconvex cases and the imposition of manifold level constraints that NRM is unable to handle with ease [25]. The method is particularly effective for managing high dimensional cases [51], including noninstantaneous flow wells and has also been extended to treat both gas-lift and choke control in each well using a mixed-integer formulation [52].

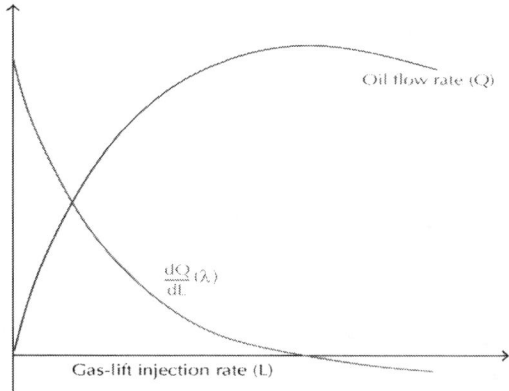

Figure 5: Lift Performance Curve and Derivative Profile.

Kosmidis et al. [53] presented a mixed-integer non-linear program (MINLP) formulation for simultaneously treating well rates and GLIR for production optimization over multiple constraints using an iterative mixed-integer linear programming (MILP) scheme. The integer variables determined well activation status and the continuous variables the amount of lift gas received by each well. The well curves were assumed to be piece-wise linear. The method accounted for the well interactions and was reported as being able to handle non-smooth and non-instantaneous flow wells. A detailed production model was employed based on a black oil reservoir model (using the Peaceman inflow performance relationship [54]), a multi-phase well-bore model, a choke model [55], and a facility gathering model [53]. The MILP approach was defined using separable programming, special ordered sets, and the adjacency assumption, similar to the formulation adopted by Handley-Schachler et al. [41] for the SLP approach. The CPLEX solver from ILOG was used for the solution of the MILP sub-problem [56]. Two test cases were examined comprising 10 naturally flowing wells and 13 gas-lifted wells. The simplification arising from linear assumptions, both for the objective function and the constraints, can be a limitation of such an approach. However, results compared favorably to traditional heuristic approaches.

Camponogara and Nakashima [57] used a dynamic-programming-(DP-) based formulation for the well-rate and lift gas allocation problem. The problem of determining which wells to produce and how much gas to allocate is a MINLP problem (as previously considered by Kosmidis et al. [53] and more recently, in Rashid et al. [52]). However, the DP algorithm solves the discretized gas-lift optimization problem approximately using precedence constraints to determine the activation of wells. Results were presented for a number of cases varying from 6 to 48 wells. The DP algorithm is relatively fast and despite the approximate nature of the approach, provides near-optimal solution for midsized networks (10–20 wells) with a sufficiently large model discretization. In general, the proposed formulation is NP-hard (this refers to non-deterministic polynomial time hard algorithm complexity) which indicates that the problem cannot be solved efficiently with increasing dimensionality due to the complexity arising from the increasing size of the connectivity graph.

The MINLP formulation for well-rate and lift gas allocation in Kosmidis et al. [53] was extended to accommodate the well scheduling problem [58]. The connection of wells to particular manifolds was treated simultaneously with well-rate management and lift gas allocation. The local solution method was deemed better than rule-based schemes, namely, those employing rules to rank and select well activation. However, the complexity of solving a large composite model for a local solution cannot guarantee a good solution overall unless executed from a good starting point or, at a higher cost, from many starting points.

NETWORK- AND RESERVOIR- BASED SOLUTIONS

Davidson and Beckner [59] used the SQP method for well-rate allocation with a reservoir simulator. A simple reservoir with 3 production wells was considered with specified production target rates.

Wang et al. [60] used the SQP solver to simultaneously solve the well-rate and lift gas allocation problem for production optimization under many constraints, including minimum and maximum flow rates, pressure limits on wells or nodes, along with available lift gas and water production limits. The method was compared to an earlier MILP plus GA-based approach by the same authors [60]. The MILP formulation excluded the back pressure effect by arbitrarily estimating the WHP in building the gas-lift performance curve. A deliverability constraint is included in the revised approach to help determine the activation state of wells and to account for the flow interaction between them. Networks of 2, 10, and 50 wells were compared with water only, water and gas, or no constraints imposed. The SQP solver was shown to be fast and robust. The MILP approach, like other simplified approaches, ignored the effect of coupling and the constraints imposed by the network. As a downside, the proposed SQP-based method does not reportedly work with looped networks due to the greater potential of flip-flopping in the solution process.

The performance of three types of solver were compared by Fujii and Horne [61] for general application to a system for production optimization: a Newton-based method, a GA, and a polytope method. The latter is a derivative-free method that uses a fixed set of points describing a polytope in the search space. The points undergo operations of expansion, contraction and reflection in pursuit of a local optimum. The method is alternatively known as the Downhill Simplex method by Nelder and Meade [62]. Two test cases (of 2 and 10 wells) were considered, with the tubing size in each well as the decision variable, although the lift gas rate could equally have been used. The polytope was effective for small problems and the GA was effective at finding the global solution, even with greater number of variables present. The quasi-Newton method became trapped in local minimum and required costly derivative evaluations. For these reasons, the method proved less suited to problems with discontinuous objective functions (see Figure 4) and comprising many variables.

Carroll and Horne [63] compared a modified Newton method with the Polytope method. A reservoir component was included in the model and an economic cost function based on present value defined. A two-variable model comprising tubing size and separator pressure was considered. The benefits of simultaneous multivariate optimization were shown over a time-stepped simulation with changing reservoir conditions. The non-smoothness of the observed noisy function can complicate algorithm convergence and the finite-difference-based derivative evaluation scheme. The second-order Newton method required evaluation of both Jacobian and Hessian matrices. In order to obtain meaningful derivatives, a large finite-difference step was suggested. However, this can affect both the performance and quality of the solution. In general, the derivative-free polytope method proved more efficient and successful at finding good solutions. Performance metrics, in the form of objective function evaluation or time cost, however, were not provided.

Palke and Horne [64] considered the application of Newton, GA and Polytope algorithms in order to optimize NPV as a function of the well-bore configuration. Control variables included tubing diameter, separator pressure, choke diameter and the depth and rate of lift gas injection. The coupled simulation considered a reservoir with a single gas-lifted well comprising reservoir, well, choke and separator component models. The Newton method (with Marquardt extension), the polytope method and the GA were compared using three problems. The first two problems concerned the tubing diameter and separator pressure, like Carroll and Horne [63], with two different fluid compositions in the well. As Fujii and Horne [61] and Carroll and Horne [63] also concluded, the use of numerical derivatives over a non-smooth surface complicates the application of gradient-based methods. The results from the Newton method were expensive and poor. The third problem examined the variation of tubing diameter and lift gas rate over 4 time steps. The polytope method did not always converge and could not easily handle problems of many variables. The GA was robust and stable but required a great number of function evaluations at significant time and computational cost. Parameter tuning was attempted to increase convergence efficiency by varying mutation rate, changing the cross-over scheme, employing fitness scaling and undertaking population culling [64]. Hepguler et al. [65], Kosmala et al. [66] and Wang and Litvak [67] all considered the integration of surface facilities with a reservoir simulator to better account for the reservoir effects over a time-lapsed simulation study. The latter employed an approximate iterative scheme using heuristics (referred to as the GLINC method) while the first two used the de-facto SQP solver.

Hepguler et al. [65] considered the variation of GLR for a model with 10 producers and 10 injectors over a 1300-day production period. Kosmala et al. [66] considered a coupled reservoir and network system for well management using downhole flow control valves and also for optimal gas allocation. The implicit SQP solver in the commercial network simulator (GAP) was employed [6]. The authors note that although the cost of coupling can be high, the results are more accurate as they take account of the limits imposed by the network model over the reservoir time-steps.

Wang and Litvak [67] introduced a multi-objective condition (besides maximizing production) to minimize the lift gas rate variation between iterations using a damping factor. This had the effect of reducing the level of fluctuations encountered. The proposed method, GLINC, was compared to previous work based on a separable programming technique using a SLP formulation, using the approach of Fang and Lo [40], together with a GA and the trust region polytope based on COBYLA method (constrained optimization by linear approximation) by Powell [37]. Although the GLINC method is easy to implement and handles flow interactions directly, it does not guarantee even a locally optimal solution due to the simplified function employed during optimization and the local nature of the search. The separable programming approach does not handle flow interactions but does return the optimal solution of the SLP problem. Comparison of the methods showed that the GLINC method was the most computationally efficient.

In the paper by Wang and Litvak [67], the long-term development plan of a North Sea field comprising 21 wells was considered for both lift gas and well-rate allocation over a period of 10,000 days. The impact of the damping factor was shown in a second test case of 18 production wells with oil, water, liquid and gas-rate constraints. While not significantly impacting cumulative oil production, the addition of the damping factor significantly reduced computational cost by reducing oscillations. Computational performance measures were provided by the authors; however, as the methods differ in their implementation, these should not be considered conclusive. It is also worth noting that the optimization procedure is called at the first few Newton iterations of each reservoir simulation time step. There is no certainty that convergence will be completed in those steps and indeed if the Newton steps are even progressing in the right direction in that period before convergence. Hence, the optimization method and the lack of damping may not be the only cause of the oscillations encountered.

INTEGRATED MODELLING APPROACH

Gutierrez et al. [25] present an integrated asset approach using decline curves for the reservoir along with network and process models [68]. The integrated asset model gives a more definitive treatment of the conditions imposed by coupling in the real world. The iterative offline-online scheme proposed by Rashid [27] was utilized for gas-lift optimization. The instabilities presented by a rigorous reservoir coupling with the production network model were avoided using predictive reservoir conditions and gas-lift allocation is performed at each coupling step. A simulation comprising 10 production wells was managed over a 20-year period for gas-lift optimization under various production constraints. These included water handling, compressor horsepower, field oil and liquid production constraints.

An integrated model comprising the reservoir, wells, surface facility model and a detailed economic model, with the purpose of optimizing the NPV of the asset, was presented by Mora et al. [69]. The model was designed to handle changes in compression, gas availability and handling capacities. The study indicated that maximum NPV does not occur when production is maximized over a common lift efficiency condition, defined as the level of production per unit gas injection over all wells, but at higher lift efficiencies. Sensitivity of the solution to oil price variation was also considered and optimization performed over a range of lift efficiencies. Results showed that depending on the price scenario selected (i.e., low to high oil price) field value is optimized by operating under different lift efficiencies. It is evident that a high oil price tends to saturate the NPV calculation. In other words, when the oil price is low, the gas and water components have a bigger impact on the NPV sensitivity to the lift efficiency employed, but if the oil price is comparatively high, these components contribute less to the NPV revenue calculation. For optimization purposes, the multi-level optimization scheme from a commercial simulator

(Landmark VIP) was employed [70]. Bieker et al. [71] presented the elements necessary for an offshore oil and gas production system for real-time optimization in a closed-loop. These include the acquisition, storage and processing of field data, simulation-model management, together with optimization and control activation procedures. The intent was to devise a system that can remotely optimize production by managing facility constraints under changing operating conditions. For example, changes in pressure, lift gas availability, separator and compressor handling limits. The daily operation was managed in a fast control loop for the short-term, while a longer-term slow loop was used to account for the changing reservoir conditions. It was noted that separate optimization levels for the slow and fast loop are necessary to handle the complexity of the composite model. Generally, the application of automatic control measures enable optimal configuration settings to be implemented with minimum intervention and time delay, helping to maximize daily production. Such schemes are particularly beneficial for remote offshore locations. Bergeron et al. [14] and Mayo et al. [36] both demonstrated these concepts in practice for a single offshore well and on a field with multiple wells, respectively.

CONCLUSIONS

In an oilfield, the daily available lift gas, often constrained due to facility conditions, is prone to variation. In addition, operating conditions and handling facilities can dictate compressor deliverability and separator limits during production, while poor allocation of the available lift gas can be economically costly, leading to over-constrained or over-designed facilities. As such, an optimal lift gas allocation is desirable to ensure that the best possible oil production or profit can be realized.

The purpose of this paper was to present a survey of the methods developed specifically to treat the continuous gas-lift optimization problem. While the basic problem concerns the optimal allocation of lift gas, the broader problem can additionally include the well-rate management problem and the well strategy problem. The former concerns the control of down-hole chokes for pressure and flow-rate control, while the latter concerns the activation state or connection of wells. In some cases, the well design for gas-lift is also considered alongside the well gas-injection rate.

The problems tackled and methods employed have evolved logically, with growing computational power and confidence, stemming from single-well analysis, to lift performance based schemes, to those utilizing rigorous network solutions, all the way to full-field integrated and closed-loop configurations. The methods and techniques employed cover a spectrum ranging from simple single-variable maximization to more sophisticated mixed-integer non-linear (MINLP) optimization schemes. In the intervening, SLP, SQP, DP, MILP, GA, TS, Polytope and Newton-based methods have all been applied in some manner. In addition, the equal-slope concept (including the marginal GOR approach) has led to the application of heuristic schemes and the development of solutions based on the separable and non-linear convex nature of the lift performance curves. Clearly, some optimization methods can be applied more readily than others depending on the formulation adopted, the solver selected and the availability of derivative information from the model. Derivative-free schemes, such as GA, TS and Polytope can be applied in most settings, but suffer from a high computational overhead if the function is costly to evaluate. On the other hand, the effectiveness of gradient-based solvers is reduced if the function is non smooth or numerical derivatives are required from an expensive simulation case, as evidenced by most coupled and integrated models. Note that the use of adjoint schemes to elicit gradient information is possible, but will be limited to a single simulation model as sensitivity information from a coupled or integrated model must be obtained collectively, most likely, by numerical means.

Thus, while some methods are more robust and extensible to large production constrained fields comprising hundreds of wells with a great number and type of variables, others are clearly limited. Classical single-well analysis neglects the interaction of other wells in a field, curve-based models neglect the back-pressure effects imposed by connected wells and network-based simulations neglect the impact of the reservoir and the process facility model. For the latter full-field simulation, the foregoing schemes are critical to ensure speed, stability and versatility of an optimization solution in real time over the fast inner loop, while the slow outer loop can be accommodated by time-stepping the inner loop. In other words, in this case, successful control and optimization of the inner loop is a necessary prerequisite in the composite full-field integrated solution. Hence, the methods devised to ensure this must be able to handle large fields, tackle difficult and non-instantaneously flowing wells and provide accurate near-optimal solutions in a reasonable time under many operating constraints. That is, they should be able to provide solutions over the period in which a well is monitored without intervention and in which operating conditions are unlikely to change significantly. In this regard, a hierarchical optimization approach may be preferential rather than an all-in-one approach. Note that when a coupled simulation includes the reservoir model not based on decline curves, the rigor of the coupling scheme, and indeed any up scaling procedure adopted, will affect the cost of the objective function evaluation and ultimately the quality of the result achieved. However, as in any enterprise, the model should be representative of the system of interest and robust for the purpose of practical optimization.

The alternative approach to the time-stepped procedure outlined above is to optimize the entire coupled or integrated system collectively over the simulation period of interest. Due to the cost associated with a single objective function evaluation, this is achieved using derivative-free methods, still at some cost, or more practically, using proxy-based methods. Here, a fast analytical approximation model is developed from a collection of representative samples using neural network, kriging or radial basis function methods. The approximation model is then used in place of the actual simulation model in the optimization step. Iterative proxy schemes have the benefit of further reducing the number of expensive function evaluations required by sampling only in the areas of the perceived optimum at each iteration [72, 73]. These methods have been demonstrated as robust, but are often limited to several dozen variables in dimensionality for practical reasons.

Thus, in general, the effectiveness of any approach adopted must consider the scalability of the solution method with increasing dimensionality (e.g., the objective function cost and, if required, the cost of derivative evaluation) along with the efficiency of the evaluation (e.g., the use of suitable proxy models) in order to minimize the overall computational cost. Mitigation of these factors is necessary for longer term simulation of large-scale fields for production optimization and especially when uncertainty is considered in the underlying models.

In summary, demonstration of a fully automated model, for gas-lift optimization in particular and for production optimization in general, that can handle the changing operational conditions and predict future development needs in a timely and robust manner is of value, and remains a challenging goal.

REFERENCES

1. K. Brown, "Overview of artificial lift systems," SPE Journal of Petroleum Technology, vol. 34, no. 10, pp. 2384–2396, 1982.

2. N. Nishikiori, R. A. Redner, D. R. Doty, and Z. Schmidt, "Improved method for gas lift allocation optimization," in Proceedings of the SPE Annual Technical Conference & Exhibition (ATCE '89), pp. 105–118, San Antonio, Tex, USA, October 1989.

3. J. C. Mantecon, "Gas-lift optimisation on Barrow Island, Western Australia," in Proceedings of the SPE Asia Pacific Oil & Gas Conference, pp. 237–246, February 1993.

4. H. Beggs, Production Optimization Using Nodal Analysis, Oil & Gas Consultants, 2nd edition, 2008.

5. Schlumberger Information Solutions, PipeSim Network Simulator, Schlumberger SIS, Abingdon, UK, 2007.

6. Petroleum Experts, GAP Network Simulator, Petroleum Experts, Edinburgh, UK, 2002.

7. A. Bahadori, S. Ayatollahi, and M. Moshfeghian, "Simulation and optimization of continuous gas lift system in aghajari oil field," in Proceedings of the SPE Asia Pacific Improved Oil Recovery Conference, Kuala Lumpur, Malaysia, October 2001.

8. D. Denney, "Simulation and optimization of continuous gas lift," Journal of Petroleum Technology, vol. 54, no. 5, p. 60, 2002.

9. S. Ayatollahi, A. Bahadori, and A. Moshfeghian, "Method optimises Aghajari oil field gas lift," Oil and Gas Journal, vol. 99, no. 21, pp. 47–49, 2001.

10. R. Vazquez-Roman and P. Palafox-Hernandez, "A new approach for continuous gas lift simulation and optimization," in Proceedings of the SPE Annual Technical Conference & Exhibition (ATCE '05), Dallas, Tex, USA, October 2005.

11. A. Bahadori and K. Zeidani, "Compositional model improves gas-lift optimization for Iranian oil field,"Oil and Gas Journal, vol. 104, no. 5, pp. 42–47, 2006.

12. A. Bahadori and K. Zeidani, "A new approach optimizes continuous gas lift system," World Oil, vol. 227, no. 11, pp. 45–52, 2006.

13. K. Dutta-Roy and J. Kattapuram, "New approach to gas-lift allocation optimization," in Proceedings of 67th Annual Western Regional Meeting, pp. 685–691, Long Beach, Calif, USA, June 1997.

14. T. Bergeron, A. Cooksey, and S. Reppel, "New automated continuous gas-lift system improves operational efficiency," in Proceedings of the SPE Mid-Continent Operations Symposium, Oklahoma, Okla, USA, March 1999.

15. W. Simmons, "Optimizing continuous flow gas lift wells— part1.," Petroleum Engineer, vol. 45, no. 8, pp. 46–48, 1972.

16. W. Simmons, "Optimizing continuous flow gas lift wells— part2," Petroleum Engineer, vol. 44, no. 10, pp. 68–72, 1972.

17. E. P. Kanu, J. Mach, and K. E. Brown, "Economic approach to oil production and gas allocation in continuous gas-lift," Journal of Petroleum Technology, vol. 33, no. 10, pp. 1887–1892, 1981.

18. T. Mayhill, "Simplified method for gas-lift well problem identification and diagnosis," in Proceedings of the SPE Fall AIME Meeting, Dallas, Tex, USA, October 1974.

19. G. A. Alarcón, C. F. Torres, and L. E. Gómez, "Global optimization of gas allocation to a group of wells in artificial lift using nonlinear constrained programming," Journal of Energy Resources Technology, Transactions of the ASME, vol. 124, no. 4, pp. 262–268, 2002.

20. J. Redden, T. Sherman, and J. Blann, "Optimizing gas-lift systems," in Proceedings of the SPE Fall AIME Meeting, pp. 1–13, Dallas, Tex, USA, October 1974.

21. J. Clegg, "Discussion of economic approach to oil production and gas allocation in continuous gas lift,"SPE Journal of Petroleum Technology, pp. 310–302, 1982.

22. E. P. Kanu, "Author's reply to discussion of economic approach to oil production and gas allocation in continuous gas lift," SPE Journal of Petroleum Technology, p. 519, 1982.

23. R. Stoisits, P. Scherer, and S. Schmidt, ". Gas optimization at the kuparak river field," in Proceedings of the SPE Annual Technical Conference & Exhibition (ATCE '94), pp. 35–42, New Orleans, La, USA, September 1994.

24. J. L. Weiss, W. H. Masino, G. P. Starley, and J. D. Bolling, "Large-scale facility expansion evaluations at the Kuparuk river field," in Proceedings SPE Regional Meeting, pp. 297–304, Bakersfield, Calif, USA, April 1990.

25. F. Gutierrez, A. Hallquist, M. Shippen, and K. Rashid, "A new approach to gas lift optimization using an integrated asset model," in SPE International Petroleum Technology Conference (IPTC '07), pp. 1371–1380, Dubai, UAE, December 2007.

26. K. Rashid, "A method for gas lift optimization," in Proceedings

of the SIAM Annual Meeting, Boston, Mass, USA, May 2008.

27. K. Rashid, "Optimal allocation procedure for gas-lift optimization," Industrial and Engineering Chemistry Research, vol. 49, no. 5, pp. 2286–2294, 2010.

28. Z. Schmidt, D. R. Doty, B. Agena, T. Liao, and K. E. Brown, "New developments to improve continuous-flow gas lift utilizing personal computers," in Proceedings of the SPE Annual Technical Conference & Exhibition (ATCE '90), pp. 615–630, New Orleans, La, USA, September 1990.

29. Y. Chia and S. Hussain, "Gas lift optimization efforts and challenges," in Proceedings of the SPE Asia Improved Oil Recovery Conference, pp. 2–9, Kuala Lumpur, Malaysia, October 1999.

30. Schlumberger Information Solutions, Eclipse Reference Manual, Schlumberger SIS, Abingdon, UK, 2006.

31. T. A. Everitt, "Gas-lift optimization in a large, mature GOM field," in Proceedings of the SPE Annual Technical Conference & Exhibition (ATCE '94), pp. 25–33, New Orleans, La, USA, September 1994.

32. R. Edwards, D. L. Marshall, and K. C. Wade, "Gas-lift optimization and allocation model for manifolded subsea wells," in Proceedings of the European Petroleum Conference (EUROPEC '90), pp. 535–545, The Hague, The Netherlands, October 1990.

33. A. A. Ferrer and R. Maggiolo, "Use of a computerized model in the optimization of continuous gas-lift operations," in Proceedings of the SPE Production Operations Symposium, pp. 87–96, April 1991.

34. Schlumberger Information Solutions, ,GOAL Gas-Lift Optimizer, Schlumberger SIS, Abingdon, UK, 1998.

35. R. F. Stoisits, E. C. Batesole, J. H. Champion, and D. H. Park, "Application of nonlinear adaptive modeling for rigorous representation of production facilities in reservoir simulation," in Proceedings of the SPE Annual Technical Conference & Exhibition (ATCE '92), pp. 425–434, Washington, DC, USA,

October 1992.

36. O. Mayo, F. Blanco, and J. Alvarado, "Procedure optimizes lift gas allocation," Oil & Gas Journal, vol. 97, no. 13, pp. 38–41, 1999.

37. P. Gill, W. Murray, and M. Wright, Practical Optimization, Academic Press, 12th edition, 2000.

38. R. Fletcher, Practical Methods of Optimization, John Wiley & Sons, 2nd edition, 2000.

39. K. K. Lo, Optimum Lift-Gas Allocations under Multiple Production Constraints, Society of Petroleum Engineers, 1992.

40. W. Y. Fang and K. K. Lo, "A generalized well-management scheme for reservoir simulation," SPE Reservoir Engineering, vol. 11, no. 2, pp. 116–120, 1996.

41. S. Handley-Schachler, C. McKie, and N. Quintero, "New mathematical techniques for the optimization of oil & gas production systems," in Proceedings of the SPE European Petroleum Conference (EUROPEC '00), pp. 429–436, October 2000.

42. R. Salazar-Mendoza, "New representative curves for gas lift optimization," in Proceedings of the SPE 1st International Oil Conference, Cancun, Mexico, August 2006.

43. S. Buitrago, E. Rodriguez, and D. Espin, "Global optimization techniques in gas allocation for continuous flow gas lift systems," in Proceedings of the SPE Gas Technology Symposium, pp. 375–383, Calgary, Canada, May 1996.

44. T. Ray and R. Sarker, "Multiobjective evolutionary approach to the solution of gas lift optimization problems," in Proceedings of the IEEE Congress on Evolutionary Computation (CEC '06), pp. 3182–3188, British Columbia, Canada, July 2006.

45. E. R. Martinez, W. J. Moreno, J. A. Moreno, and R. Maggiolo, "Application of genetic algorithm on the distribution of gas lift injection," in Proceedings of the 3rd SPE Latin American and Caribbean Petroleum Engineering Conference, pp. 811–818, Buenos Aires, Argentina, April 1994.

46. K. Deb, Multi-Objective Optimization Using Evolutionary Algorithms, John Wiley & Sons, 2001.

47. K. Dutta-Roy, S. Barua, and A. Heiba, "Computer-aided gas field planning and optimization," inProceedings of the SPE Production Operations Symposium, pp. 511–515, Oklahoma, Okla, USA, March 1997.

48. M. S. Nadar, T. S. Schneider, K. L. Jackson, C. J. N. McKie, and J. Hamid, "Implementation of a total-system production-optimization model in a complex gas lifted offshore operation," SPE Production and Operations, vol. 23, no. 1, pp. 5–13, 2008.

49. M. Vazquez, A. Suarez, H. Aponte, L. Ocanto, and J. Fernandes, "Global optimization of oil production systems, a unified operational view," in Proceedings of the SPE Annual Technical Conference & Exhibition (ATCE '01), New Orleans, La, USA, October 2001.

50. R. Stoisits, D. MacAllister, M. McCormack, A. Lawal, and D. Ogbe, "Production optimization at the kuparak river field utilizing neural networks and genetic algorithms," in Proceedings of the SPE Mid-Continent Operations Symposium, Oklahoma, Okla, USA, March 1999.

51. S. Moitra, S. Chand, S. Barua, D. Adenusi, and V. Agrawal, "A field-wide integrated production model and asset management system for the mumbai high field," in Proceedings of the SPE Offshore Technology Conference, Houston, Tex, USA, April 2007.

52. K. Rashid, S. Demirel, and B. Couët, "Gas-lift optimization with choke control using a mixed-integer nonlinear formulation," Industrial and Engineering Chemistry Research, vol. 50, no. 5, pp. 2971–2980, 2011.

53. V. D. Kosmidis, J. D. Perkins, and E. N. Pistikopoulos, "Optimization of well oil rate allocations in petroleum fields," Industrial and Engineering Chemistry Research, vol. 43, no. 14, pp. 3513–3527, 2004.

54. D. W. Peaceman, "Interpretation of well-block pressures

in numerical reservoir simulation," Society of Petroleum Engineers, vol. 18, no. 3, pp. 183–194, 1978.

55. R. Sachdeva, Z. Schmidt, J. P. Brill, and R. M. Blais, "Two pahse flow through chokes," in SPE Annual Technical Conference and Exhibition, Society of Petroleum Engineers, New Orleans, Louisiana, 5-8 October 1986.

56. ILOG, ILOG CPLEX User's Manual, ILOG, 2002.

57. E. Camponogara and P. H. R. Nakashima, "Solving a gas-lift optimization problem by dynamic programming," European Journal of Operational Research, vol. 174, no. 2, pp. 1220–1246, 2006.

58. V. D. Kosmidis, J. D. Perkins, and E. N. Pistikopoulos, "A mixed integer optimization formulation for the well scheduling problem on petroleum fields," Computers and Chemical Engineering, vol. 29, no. 7, pp. 1523–1541, 2005.

59. J. E. Davidson and B. Beckner, "Integrated optimization for rate allocation in reservoir simulation," inSPE Reservoir Simulation Symposium, Houston, Tex, USA, February 2003.

60. P. Wang, M. Litvak, and K. Aziz, "Optimization of production operations in petroleum fields.," inProceedings of the SPE Annual Technical Conference & Exhibition (ATCE '02), San Antonio, Tex, USA, September 2002.

61. H. Fujii and R. Horne, "Multivariate optimization of networked production systems," SPE Production & Facilities, vol. 10, no. 3, pp. 165–171, 1995.

62. W. Press, S. Teukolsky, W. Vetterling, and B. Flannery, Numerical Recipes in C++, Cambridge University Press, Cambridge, UK, 3rd edition, 2002.

63. J. Carroll and R. Horne, "Multivariate optimization of production systems," SPE Journal of Petroleum Technology, vol. 44, no. 7, pp. 782–789, 1992.

64. M. R. Palke and R. N. Horne, "Nonlinear optimization of well production considering gas lift and phase behavior," in Proceedings of the SPE Production Operations Symposium, pp. 341–356, Oklahoma, Okla, USA, March 1997.

65. G. G. Hepguler, K. Dutta-Roy, and W. A. Bard, "Integration of a field surface and production network with a reservoir simulator," SPE Computer Applications, vol. 12, no. 4, pp. 88–92, 1997.

66. A. Kosmala, S. I. Aanonsen, A. Gajraj et al., "Coupling of a surface network with reservoir simulation," in Proceedings of the SPE Annual Technical Conference & Exhibition (ATCE '03), pp. 1477–1487, Denver, Colo, USA, October 2003.

67. P. Wang and M. Litvak, "Gas lift optimization for long-term reservoir simulations," SPE Reservoir Evaluation and Engineering, vol. 11, no. 1, pp. 147–153, 2008.

68. Schlumberger Information Solutions, Avocet-IAM Integrated Asset Modeller, Schlumberger SIS, Calgary, Canada, 2007.

69. O. Mora, R. A. Startzman, and L. Saputelli, "Maximizing net present value in mature gas-lift fields," inProceedings of the SPE Hydrocarbon Economics and Evaluation Symposium: Hydrocarbon Development—A Global Challenge, pp. 123–131, Dallas, Tex, USA, April 2005.

70. Landmark Graphics Corporation, VIP Simulator, Landmark Graphics Corporation, 2003.

71. H. P. Bieker, O. Slupphaug, and T. A. Johansen, "Real-time production optimization of oil and gas production systems: a technology survey," SPE Production and Operations, vol. 22, no. 4, pp. 382–391, 2007.

72. B. Couët, H. Djikpesse, D. Wilkinson, and T. Tonkin, "Production enhancement through integrated asset modeling optimization," in SPE Production and Operations Conference and Exhibition (POCE '10), pp. 375–384, Tunis, Tunisia, June 2010.

73. K. Rashid, S. Ambani, and E. Cetinkaya, "An adaptive multiquadric radial basis function method for expensive black-box mixed-integer nonlinear constrained optimization," Engineering Optimization. In press.

Chapter 2

Gas Production from a Cold, Stratigraphically-Bounded Gas Hydrate Deposit at the Mount Elbert Gas Hydrate Stratigraphic Test Well, Alaska North Slope: Implications of Uncertainties

G.J. Moridis[a], S. Silpngarmlert[b], M.T. Reagan[a], T. Collett[c], and K. Zhang[a]

[a]Lawrence Berkeley National Laboratory, 1 Cyclotron Rd., Berkeley, CA 94720, United States

[b]ConocoPhillips, Houston, TX 77252, USA

cU.S. Geological Survey, Denver Federal Center, MS-939, Denver, CO 80225, USA

ABSTRACT

As part of an effort to identify suitable targets for a planned long-term field test, we investigate by means of numerical simulation the gas productionpotential from unit D, a stratigraphically bounded (Class 3) permafrost-associated hydrate occurrence penetrated in the BPXA-DOE-USGS Mount Elbert Gas Hydrate Stratigraphic Test Well on North Slope, Alaska. This shallow, low-pressure deposit has high porosities (= 0.4), high intrinsic permeabilities ($k = 10^{-12}$ m^2) and high hydrate saturations ($S_H = 0.65$). It has a low temperature ($T = 2.3–2.6$ °C) because of its proximity to the overlying permafrost. The simulation results indicate that vertical wells operating at a constant bottomhole pressure would produce at very low rates for a very long period. Horizontal wells increase gas production by almost two orders of magnitude, but production remains low. Sensitivity analysis indicates that the initial deposit temperature is by the far the most important factor determining production performance (and the most effective criterion for target selection) because it controls the sensible heat available to fuel dissociation. Thus, a 1 °C increase in temperature is sufficient to increase the production rate by a factor of almost 8. Production also increases with a decreasing hydrate saturation (because of a larger effective permeability for a given k), and is favored (to a lesser extent) by anisotropy.

INTRODUCTION

Background

Gas hydrates are solid crystalline compounds in which gas molecules (referred to as guests) occupy the lattices of ice crystal structures (called hosts). Their formation and dissociation is described by the general equation

$$G + N_H\, H_2O = G{\cdot}N_H\, H_2O,$$

where N_H is the hydration number, and G is a hydrate-forming gas. Natural hydrates in geological systems contain $G = CH_4$ as their main gas ingredient, and occur in two distinctly different geologic settings where the necessary conditions of low T and high P exist for their formation and stability: in the permafrost and in deep ocean sediments.

Although the magnitude of the CH_4 resource trapped in hydrates is the subject of rigorous debate, and the estimates vary widely between 10^{15} and 10^{18} ST m^3 (Sloan and Koh, 2008, Milkov, 2004 and Klauda and Sandler, 2005), there is general consensus that it is huge, easily exceeding the total energy content of the known conventional fossil fuel resources. Even if only a fraction of the most conservative estimate of the resource is recoverable, the CH_4 amounts involved are sufficiently large to demand evaluation of the hydrate potential as an energy source (Makogon, 1987, Dallimore et al., 1999 and Dallimore and Collett, 2005). To that end, a global effort is currently in progress to assess the resource (Moridis et al., 2009), and the ever-increasing global energy demand, the dwindling conventional fossil hydrocarbon reserves, and the environmental desirability of CH_4 as a fuel have added to the impetus for this effort. As result, there has been a proliferation of recent studies evaluating the technical and economic feasibility of gas production from natural hydrate accumulations, e.g., Moridis, 2003 and Moridis et al., 2004; Hong and Darwish-Pooladi, 2005; Sun and Mohanty, 2005; Moridis and Sloan, 2007, Moridis and Reagan, 2007a, Moridis and Reagan, 2007b, Moridis and Reagan, 2007c, Moridis et al., 2008b, Kurihara et al., 2005 and Kurihara et al., 2008.

Gas can be produced from hydrates by inducing dissociation via one of the three main dissociation methods (Makogon, 1997) or combinations thereof: (1) depressurization below the hydration pressure P_e (as defined by the Lw–H–V and I–H–V three-phase lines in Fig. 1.1) at the temperature T, (2) thermal stimulation, based on raising T above the hydration temperature T_e at the prevailing pressure P, and (3) the use of inhibitors (such as salts and alcohols) that shift the $P_e - T_e$ equilibrium.

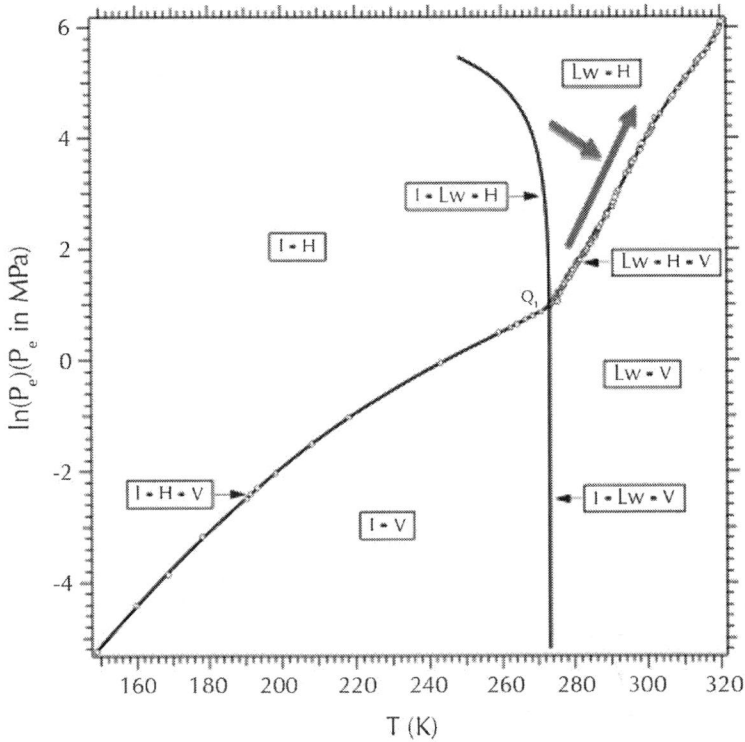

Figure.1.1: Pressure–temperature equilibrium relationship in the phase diagram of the water–CH$_4$–hydrate system (Moridis, 2003), Lw: Liquid water; H: Hydrate; V: Vapor (gas phase); I: Ice; Q$_1$: Quadruple point (=I + Lw + H + V). The two arrows show the direction of increasing thermodynamic desirability of a deposit as a production target.

Objectives and Approach

This investigation is part of an effort led by the U.S. Department of Energy to identify appropriate targets for a long-term field test of production from permafrost-associated hydrate deposits (Boswell et al., 2008). The main objectives of this study are (a) to evaluate the gas production potential of the unit D hydrate accumulation at the Mount Elbert site, North Slope, Alaska, and, should this be deemed unsatisfactory, (b) to determine through sensitivity analysis the conditions and properties that can serve as criteria to identify other deposits as suitable candidate for a successful field test of production.

Unit D at the Mount Elbert site (described in more detail in Section 2) is a relatively shallow deposit that is cold (2.3–2.6 °C) because of its proximity to the permafrost. It is a typical Class 3 deposit, i.e., it involves a single zone – the hydrate-bearing layer (HBL), confined by near-impermeable top and bottom boundaries – and is characterized by the absence of an underlying zone of mobile fluids. As discussed in detail byMoridis and Reagan, 2007a and Moridis and Reagan, 2007b, depressurization appears to be the production method of choice because of its simplicity, its technical and economic effectiveness, the fast response of hydrates to the rapidly propagating pressure wave, the near-incompressibility of water, and the large heat capacity of water. The latter plays a significant role in providing part of the heat needed to support the strongly endothermic dissociation reaction.

Because of the high initial hydrate saturation S_H in the HBL, the effective permeability k_{eff} is very low and constant-rate production is not feasible, while pure thermal stimulation is an unattractive option because of its limited effectiveness for reasons discussed in detail by Moridis and Reagan (2007a). Thus, our studies focused exclusively on production under a constant bottomhole pressure P_w regime because earlier studies (Moridis and Reagan, 2007a and Reagan et al., 2008) had indicated this to be a promising (and possibly the only) option in production from Class 3 deposits of similar attributes. We investigated the performance of both vertical and horizontal wells, and we conducted a sensitivity analysis to determine the most important factors affecting production.

THE MOUNT ELBERT SITE

Regional Geological System Description

The geology and petroleum geochemistry of the rocks on the North Slope of Alaska where gas hydrates are encountered are described in considerable detail in a number of publications (Bird and Magoon, 1987 and Collett, 1993). The first direct confirmation of gas hydrate on the North Slope was provided by data from a single well (the Northwest Eileen State-2 well, located in the northwest part of the Prudhoe Bay Field), in which studies of pressurized core samples, downhole logs, and production testing had confirmed the occurrence of three gas-hydrate-bearing stratigraphic units (Collett, 1993). Analysis of downhole log data from an additional 50 exploratory and production wells in the same area provided additional indications of hydrate occurrence in six laterally continuous sandstone and conglomerate units (A–F), which are all confined to the geographical area shown in Fig. 2.1. Collett (2007) indicated that the hydrate units appear to trap down-dip several large free-gas accumulations (Fig. 2.1; units A through D). The volume of gas within the Eileen Gas Hydrate Accumulation (Collett, 2007) is estimated at about twice the volume of known conventional gas in the Prudhoe Bay Field (Collett, 1993), and ranges between 1.0×10^{12} and 1.2×10^{12} m^3 STP (Collett, 2007).

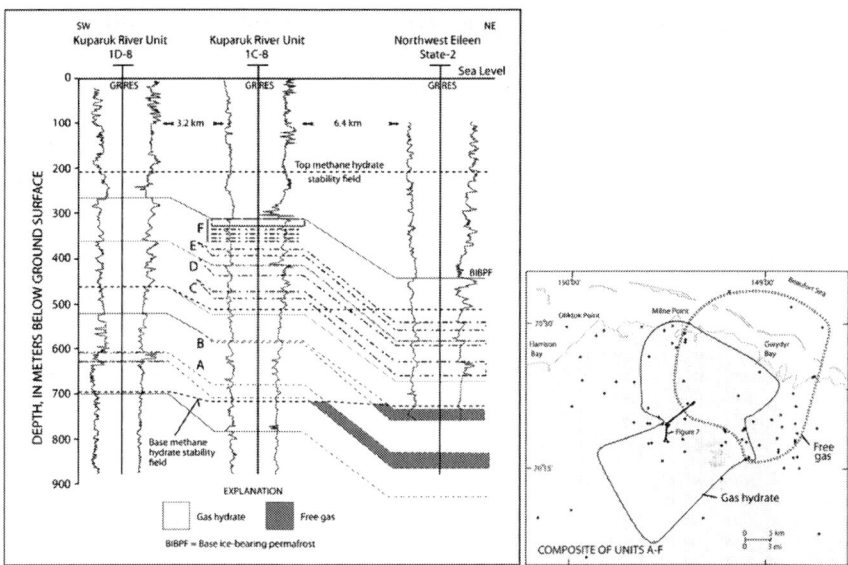

Figure. 2.1: (a) Cross section showing the lateral and vertical extent of gas hydrates and underlying free-gas occurrences in the Prudhoe Bay–Kuparuk River area in northern Alaska. See Fig. 2.1b for location of cross section. The gas-hydrate-bearing units are identified with the reference letters A through F (Collett, 1993), and their positions relative to the permafrost and to the base of the hydrate stability zone are shown; (b) Composite map of all six gas-hydrate/free-gas units (units A–F) from the Prudhoe Bay–Kuparuk River area in northern Alaska (Collett, 1993).

Previous Studies

Previous and current studies of gas production from hydrates in the North Slope of Alaska involve collaborations that are spearheaded by BP Exploration (Alaska – BPXA), Inc., the U.S. Department of Energy, and the U.S. Geological Survey, and involve several other organizations. This effort is supported by the Methane Hydrate Research and Development Act (enacted by the U.S. Congress in 2000 and renewed in 2005), and aims to determine the viability of the North Slope hydrates as an energy source (Hunter et al., 2011) through investigations that will culminate with a long-term

(1.5–2 years) field test of gas production (Boswell et al., 2008). Analysis of geophysical surveys and well log data led the team to the installation of a well in 2007 at a previously undrilled, fault-bounded accumulation named the "Mount Elbert" prospect to acquire critical reservoir data needed to develop a longer-term production test program. The BPXA-DOE-USGS Mount Elbert Gas Hydrate Stratigraphic Test Well (the Mount Elbert Well) was drilled to a depth of 915 m using chilled oil-based drilling fluid to avoid the inhibitor-induced dissociation caused by the salts and alcohols in conventional muds. A remarkable achievement was the recovery of significant lengths of core from the hydrate intervals, which were used for subsequent analyses of pore water geochemistry, microbiology, gas chemistry, petrophysical properties, and thermal and physical properties. After a battery of well log surveys was completed, a Schlumberger Modular Dynamic Testing (MDT) was conducted in two reservoir-quality sandy hydrate-bearing sections with high S_H (60–75%). Gas was produced from the gas hydrates in each of the tests. This study has yielded one of the most comprehensive datasets yet compiled on a naturally occurring gas hydrate geologic deposit (Collett, 2007).

Extensive discussions of the Mount Elbert geology and analyses of the various tests conducted at the site are presented in various papers in this volume (Boswell et al., 2011).

The Unit D Hydrate Deposit

Fig. 2.2 shows units C and D at the Mount Elbert site, and Fig. 2.1 shows their location relative to (a) the permafrost and (b) the predicted base of the methane hydrate stability zone. Unit D is a shallow permafrost-associated hydrate deposit, with a HBL beginning at a depth of $z = -616.6$ m. The deposit is about 11.3 m thick, is bounded by nearly impermeable shale layers, and has high porosity, permeability and hydrate saturation (Winters et al., 2011). Because of its proximity to the permafrost, its temperature is low, ranging between $T_T = 2.3$ °C and $T_B = 2.6$ °C at the HBL top and bottom, respectively. The pressure at the HBL top is a low $P_T =$

6.386 MPa. The properties and initial conditions of the unit D and its boundaries are listed in Table 1.

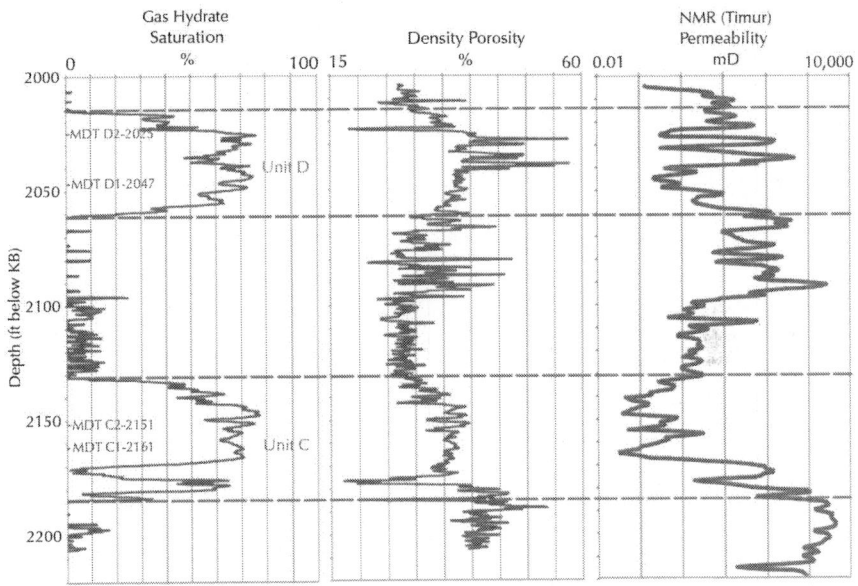

Figure. 2.2: Relation of the C and D units at the Mount Elbert-01 well. The locations of the MDT tests in the two units are also shown.

Table 1: Hydrate Deposit Properties in unit D, Mount Elbert Site

Parameter	Value
Hydrate zone thickness	11.3 m
Initial pressure at top of HBL (P_T)	6.386 MPa
Initial temperature at top of HBL (T_T)	2.3 °C
Initial temperature at base of HBL (T_B)	2.6 °C
Gas composition	100% CH_4
Initial saturations in the HBL	$S_H = 0.65$, $S_A = 0.35$
Intrinsic permeability of HBL $k_r = k_x = k_z$	10^{-12} m^2 (=1 D)
Porosity of HBL $\varnothing$	0.4
Compressibility of HBL	5×10^{-9} Pa^{-1}

Intrinsic permeability $k_r = k_x = k_z$ (overburden & underburden)	0 m² (=0 D)
Porosity of overburden & underburden	0.005
Grain density ρ_R (all formations)	2750 kg/m³
Constant bottomhole pressure (P_w)	3 MPa
Dry thermal conductivity ($k_{\theta RD}$) (all formations)	0.5 W/m/K
Wet thermal conductivity ($k_{\theta RW}$) (all formations)	3.1 W/m/K
Composite thermal conductivity model (Moridis et al., 2008c)	$k_{\theta C} = k_{\theta RD} + (S_A^{1/2} + S_H^{1/2})(k_{\theta RW} - k_{\theta RD}) + \square S_l k_{\theta I}$
Capillary pressure model (van Genuchten, 1980)	$Pcap = -P_0[(S^*)^{-1/\lambda} - 1]^{-\lambda}$ $$S^* = \frac{(S_A - S_{irA})}{(S_{mxA} - S_{irA})}$$
S_{irA}	1
λ (White, 2008)	0.77437
P_0 (White, 2008)	5 × 10³ Pa
Relative permeability model (Moridis et al., 2008c)	$k_{rA} = (S_A^*)^n$ $k_{rG} = (S_G^*)^m$ $S_A^* = (S_A - S_{irA})/(1 - S_{irA})$ $S_G^* = (S_G - S_{irG})/(1 - S_{irA})$ EPM model
n; m (from Anderson et al., 2008 and Anderson, et al., 2011b)	4.2; 2.5
S_{irG}	0.02
S_{irA}	0.20

In terms of desirability as a production target for a long-term production test, our initial perception of the advantage of unit D over unit C was that it is a Class 3 deposit, i.e., it is characterized by the absence of an underlying zone of mobile fluids, as opposed to unit C, which is connected to a deep, extensive aquifer that makes depressurization challenging and water disposal an additional complication. Continuing studies have provided indications that unit D is likely in communication with some underlying water-bearing sand sections, but the extent of this communication is unknown. For the purpose of this study, the production modeling is based on the assumption that unit D is a Class 3 deposit with no connection to a underlying zone of mobile fluids.

Compared to unit D, unit C is thicker and warmer by about 1°C (Collett et al. 2011a). Although this may initially appear unimportant, the small increase in temperature can make a very significant difference in production from hydrates because it increases the sensible heat that is available to support the endothermic hydrate dissociation: the lower the initial temperature T, the bigger the potential effect of an additional 1 °C on dissociation and gas production.

Units C and D have similar properties, and similar S_H. Because of the MDT test that was conducted within the C unit, it was possible to determine some of its in-situ properties by history matching the MDT data (Anderson et al., 2011b). Other unit C and unit D properties and conditions were determined from well log analyses and core studies of samples retrieved during drilling (Collett et al. 2011b).

THE NUMERICAL MODELS AND SIMULATION APPROACH

The Numerical Simulation Code

We used the TOUGH + HYDRATE simulator (Moridis et al., 2008c and Zhang and Moridis, 2008) to conduct the numerical studies in this paper. This code (hereafter referred to as $T + H$) can model all the known processes involved in the system response of natural CH_4-hydrates in complex geologic media, including the flow of fluids and heat, the thermophysical properties of reservoir fluids, thermodynamic changes and phase behavior, and the non-isothermal chemical reaction of CH_4-hydrate formation and/or dissociation, which can be described by either an equilibrium or a kinetic model (Kim et al., 1987 and Clarke and Bishnoi, 2000; Kowalsky and Moridis, 2007). $T + H$ is a compositional simulator, and its formulation accounts for heat and up to four mass components (i.e., H_2O, CH_4, CH_4-hydrate, and water-soluble inhibitors such as salts or alcohols) that are partitioned among four possible phases: gas, aqueous liquid, ice, and hydrate. The $T + H$ code can describe all the 15 possible thermodynamic states (phase combinations) of the $CH_4 + H_2O$ system and any combination of the three hydrate dissociation methods. It can handle the phase changes, state transitions, strong nonlinearities and steep solution surfaces that are typical of hydrate dissociation problems. Because of the very large computational requirements of this type of problem and the use of very large grids (see Section 3.3), we used the distributed-memory, massively parallel version of the code (Zhang and Moridis, 2008) in the simulations discussed in this paper.

System Geometry

The geologic system in this study corresponds to a location at the Mount Elbert site where the top of the HBL is at a depth of $z = -616.6$ m. This is a typical Class 3 deposit, in which the 11.3-m-thick HBL is overlain and underlain by nearly impermeable boundaries, i.e., shale strata. Based on experienced gained in earlier studies (Moridis and Reagan, 2007a, Moridis and Reagan, 2007b and Moridis et al., 2008b) and preliminary scoping calculations, the simulation domain was extended 30 m into the overburden and underburden of the HBL, a distance that was deemed sufficient to allow accurate heat exchange with the deposit during the production period.

We investigated the performance of both a single vertical and a single horizontal well producing from sections (cylindrical and rectangular) of the same hydrate deposit, using the same surface area and including the same hydrate volume in each simulation configuration. The outer radius of the cylindrical section was $r_{max} = 400$ m, corresponding to a well spacing of 50 ha (125 acres). The rectangular section with the same area and hydrate volume had a square footprint with a side $L_y = 709$ m. The geometry and well configuration of these two Class 3 systems are shown in Fig. 3.1 and Fig. 3.2. The horizontal well was placed at the top of the HBL to capitalize on gas buoyancy and accumulation at this location, in addition to minimizing water production ($Z_w = 0$, see Fig. 3.2). Both the vertical and the horizontal well had a radius $r_w = 0.1$ m.

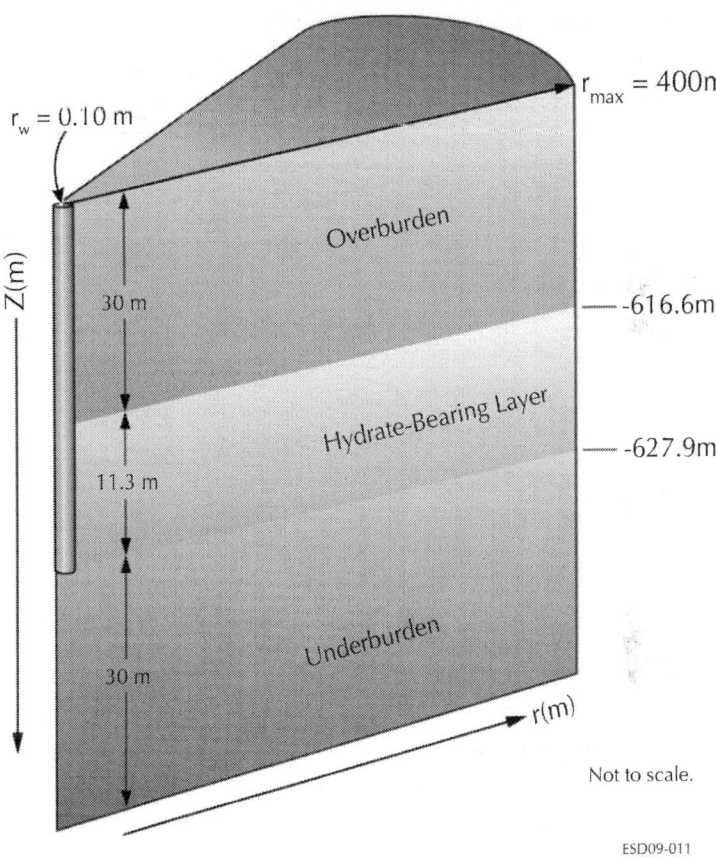

Figure. 3.1: System geometry and configuration of the single vertical well producing from a cylindrical section of the Unit D Class 3 hydrate deposit at the Mount Elbert site.

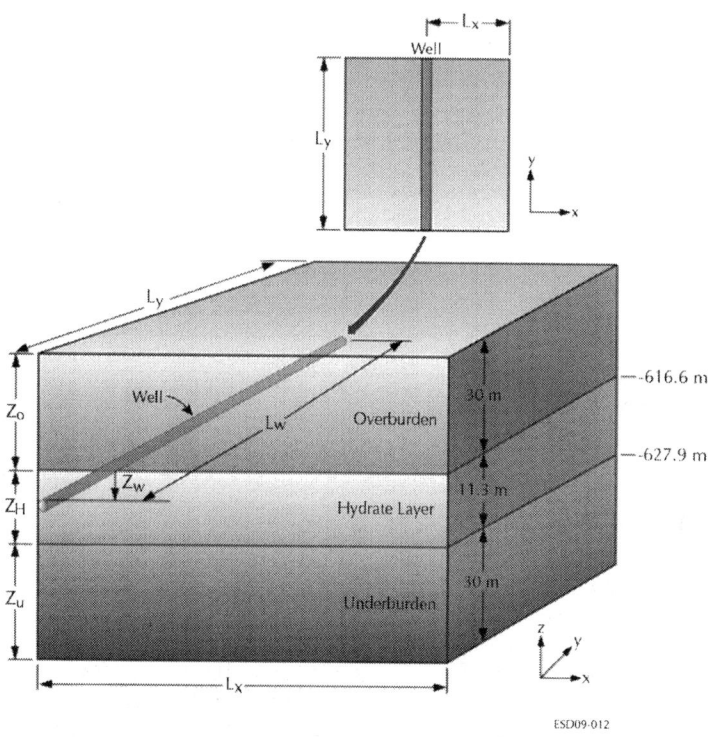

ESD09-012

Figure 3.2: System geometry and configuration of the horizontal well producing from a rectangular section of the Unit D hydrate deposit that has the same area and hydrate volume as the cylindrical section of Fig. 3.1 Note that $L_w = L_y = 709$ m, $L_x = L_y/2 = 354.5$ m, $Z_w = 0$ m.

Domain Discretization

For maximum accuracy, very fine grids were used in the simulation of production from both the cylindrical and rectangular sections of the hydrate deposit. The cylindrical domain of Fig. 3.1 was discretized into 200 × 300 = 60,000 gridblocks in (r,z), resulting in a system of 240,000 equations. Discretization along the radial direction was non-uniform, increasing logarithmically from r_w to r_{max}, with $r_0 = 0.05$ m. Discretization along the z-axis was uniform (with $z = 0.1$ m) within the HBL and its immediate vicinity, but non-uniform (with z increasing) near the top and bottom of the domain.

In the study of the performance of the horizontal well, we used only a single slice of unit thickness on the (x,z) plane, i.e., perpendicular to the horizontal well (Fig. 3.2). Implicit in this approach is the assumption of uniformity along the well length L_w, i.e., along the y-axis. While this assumption may not be always valid in light of expected pressure variations along the length of the well, it is a good first-order approximation, it can be used to bound the expected solution through the choice of an appropriate range of well pressures in the studied slices, and it allows high-definition in the description of the system behavior without resulting in a prohibitively large grid. As in the case of production from a single vertical well in a cylindrical section of the hydrate deposit, the 2D domain in (x,z) was discretized into $200 \times 300 = 60,000$ gridblocks in (x,z). The vertical discretization was the same as in the case of the cylindrical system. Discretization along x-axis was non-uniform, increasing logarithmically from $x_0 = r_w$ to L_x, with $x_0 = 0.05$ m.

Such a fine discretization is important (and possibly necessary) for accurate predictions when solid phases such as ice and hydrates are involved (Moridis and Reagan, 2007a, Moridis and Reagan, 2007b and Moridis and Reagan, 2007c). This high degree of refinement provided the level of detail needed to capture important processes near the wellbore and in the entire hydrate-bearing zone. Assuming an equilibrium reaction of hydrate dissociation during this long-term production process (Kowalsky and Moridis, 2007), and accounting for the water salinity, the grid resulted in 240,000 coupled equations that were solved simultaneously.

System Properties and Well Description

The hydraulic and thermal properties of the various geological media (the HBL and the confining layers) in unit D, as well as the initial conditions, were obtained from data based on the first field test at the site (Anderson et al., 2008 and Anderson et al., 2011a), and are listed in Table 1. We assumed that the initial hydrate and aqueous saturations (S_H and S_A, respectively) were uniformly distributed in the HBL, and that the overburden and underburden had both the same properties. The relative permeability relationships and the corresponding parameters were based on data obtained from history matching of the results of MDT test that had been conducted at the C unit at the same site (Anderson et al., 2008 and Anderson et al., 2011a), which appeared to have similar properties. The capillary pressure relationships and parameters were determined from the particle size analysis of porous media samples from the deeper (but similar) C unit (White, 2008) and were consistent with the porosity, , and permeability, k, of the D unit.

The importance of the near-well region dictated the physical representation of the wellbore in the vertical well study. To avoid a theoretically correct but computationally intensive solution of the Navier–Stokes equation, we approximated wellbore flow by Darcian flow through a pseudo-porous medium describing the interior of the well. Earlier studies had shown the validity of this approximation (Moridis and Reagan, 2007b and Moridis and Reagan, 2007c). This pseudo-medium had $\varphi = 1$, a very high $k = 10^{-9}-10^{-8}$ m^2(=1000–10,000 Darcies), a capillary pressure $P_c = 0$, a relative permeability that was a linear function of the phase saturations in the wellbore, and a low (but nonzero) irreducible gas saturation $S_{irG} = 0.005$ (necessary to allow the emergence of a free-gas phase in the well).

Initial and Boundary Conditions

The no-flow conditions (of fluids and heat) that were applied at the reservoir outer boundaries (at a radius $r = r_{max}$ and at $x = L_x = L_y/2$, See Fig. 3.1 and Fig. 3.2) implied the presence of other wells with the same characteristics in adjacent sections of the hydrate deposit on the same spacing patterns.

We determined the initial conditions in the reservoir by following the initialization process described by Moridis and Reagan, 2007a and Moridis and Reagan, 2007b. The temperatures at the top and bottom of the HBL (T_T and T_B, respectively) have been extrapolated from high resolution equilibrated temperature log surveys in a nearby well (Collett et al. 2011b). In both the cylindrical and the rectangular systems, the uppermost and lowermost gridblock layers (i.e., at the top of the overburden and at the bottom of the underburden in the simulated domains, where $z = 0.001$ m) were treated as boundaries with constant conditions and properties. The temperatures at the upper and lower domain boundaries (T_U and T_L, respectively) were determined through a trial-and-error simulation process that resulted in the known T_P and T_B across the HBL. Note that the shales in the overburden and underburden were treated as impermeable (Table 1).

Knowing (a) the depth at the base of the HBL, and (b) assuming that the pressures in the subsurface follow the hydrostatic distribution—a hypothesis supported by field observations (Collett et al., 1988) and other observations (Wright et al., 1999) in hydrate accumulations—we determined the pressure P_T (at $z = -616.6$ m, see Fig. 3.1 and Fig. 3.2) using the P-, T- and salinity-adjusted water density (1005 kg/m³ at atmospheric pressure). Then, using P_T and the boundary temperatures T_T and T_B, the hydrostatic gradient and representative thermal conductivity values were employed to determine the P- and T-profiles in the domains by means of a short simulation.

For reasons explained in detail by Moridis and Reagan (2007b), depressurization appears to be the most effective dissociation strategy, and a constant-pressure regime (involving a constant bottomhole pressure P_w at the well) is the most promising method of gas production from Class 3 hydrate deposits. Its numerical representation involves treating the well as an internal boundary. In the case of a vertical well, this boundary is placed in the gridblock above the uppermost cell in the well. By imposing a constant P_w, a thermal conductivity $k = 0$ W/m/K, and a realistic (though unimportant) constant temperature T_w at this internal boundary, the correct constant-P condition was applied to the well while avoiding any non-physical temperature distributions in the well itself (the large advective flows into the uppermost gridblock from its immediate neighbor eliminated any unrealistic heat transfer effects that could have resulted from an incorrect k and/or T_w). In our study, the $P_w = 3.0$ MPa exceeds the pressure at the quadruple point P_Q, thus eliminating the possibility of ice formation and the corresponding potentially adverse effect on k_{eff}

Simulation Process and Outputs

The maximum simulation period was initially the typical 30-year life span of a well, but it had to be extended to 50 years in the case of the horizontal well in order to investigate its very-long-term performance. In the course of the simulation, the following conditions and parameters were monitored: Spatial distributions of P, T, and gas and hydrate phase saturations (S_G and S_H); Volumetric rate of CH_4 released from dissociation and of CH_4 production at the well (Q_R and Q_P, respectively); Cumulative volume of CH_4 released from dissociation, produced at the well, or remaining in the deposit as free gas (V_R, V_P and V_F, respectively); water mass production rate at the well (Q_W) and cumulative mass of produced water (M_W); the remaining hydrate as a fraction of its original mass ($M_{HR} = M_{H,}t/M_{H,0}$, where $M_{H,0}$ and $M_{H,}t$ are the hydrate mass in the reservoir at times 0 and t, respectively).

THE CASE OF PRODUCTION US-ING A VERTICAL WELL

Gas Production

Fig. 4.1 shows the evolution of Q_R and Q_P from the single vertical well at the center of the cylindrical reservoir of Fig. 3.1 over time. The most important conclusion from the review of Fig. 4.1 is that the CH_4 release and production remain very low for a very long period. Thus, Q_R and $Q_P < 7 \times 10^{-4}$ ST m³/s (<2000 ST ft³/s) for 8000 days, i.e., almost 22 years. After that time, both Q_R and Q_P appear to increase exponentially with time, but are lower than 2.3×10^{-3} ST m³/s (=7000 ST ft³/s) even after t = 10,800 days (30 years). The low production rate is caused by the very low initial temperature of the hydrate in the HBL. The low T reduces the rate of the dissociation reaction and severely reduces the sensible heat that is available to support it. The cumulative produced volume V_p in Fig. 4.2 provides further confirmation of the limited potential of unit D as a target for production from hydrates by depressurization: after continuous production for t = 30 years, a mere V_p = 5.3 × 10⁵ ST m³ (=1.9 × 10⁷ ST ft³) of CH_4 have been produced.

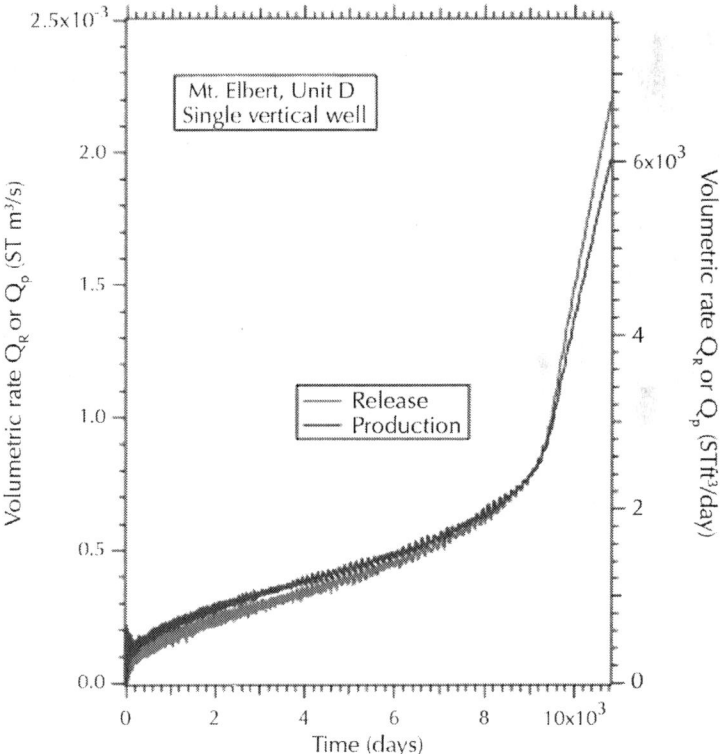

Figure. 4.1: Evolution of Q_R and Q_p during production from Unit D using a vertical well.

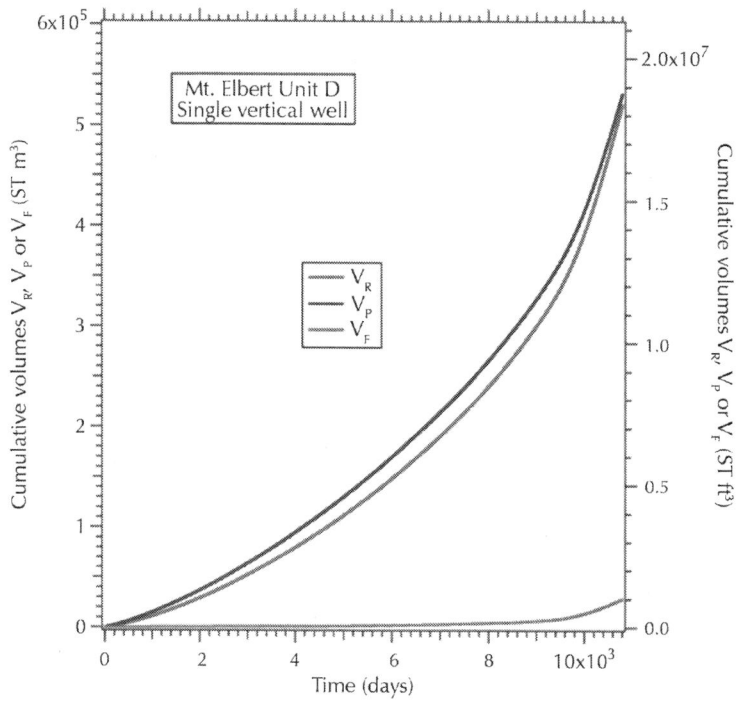

Figure. 4.2: V_R, V_P and V_F during production from unit D using a vertical well.

An interesting observation from Fig. 4.1 and Fig. 4.2 is that gas release from dissociation lags production for a very long time. Thus, $Q_P > Q_R$ for $t < 9000$ days (Fig. 4.1), and $V_P > V_R$ even at the end of the 30-year-long production period (Fig. 4.2). The source of the additional gas is dissolved CH_4 that is released from solution as the pressure in the formation drops (and the CH_4 solubility decreases) during production. Note the very low level of free gas, V_F, in the reservoir during production (Fig. 4.2), which does not exhibit an upward trend until the time of the rapid increase in Q_R and Q_P (Fig. 4.1). The low levels of V_F, and the near-parity of V_R and V_P (and Q_R and Q_P), indicate that there is little gas accumulation in the reservoir, and most of the gas released from dissociation and dissolution is produced at the vertical well. After $t = 9000$ days, we see that gas release begins to outpace gas production, indicating that hydrate dissociation has finally begun to create significant free gas in the reservoir (as depressurization has finally destabilized the cold, stable initial state of the system), allowing production to increase exponentially.

Water Production and Effectiveness of Dissociation

The water production rate Q_W in Fig. 4.3 remains at low levels and within a very narrow range (0.02 kg/s $< Q_W <$ 0.026 kg/s) during the entire 30-year production period. The relative stability of Q_W leads to the near-linear appearance of the cumulative water mass M_W curve. It is obvious that M_W is at easily manageable levels.

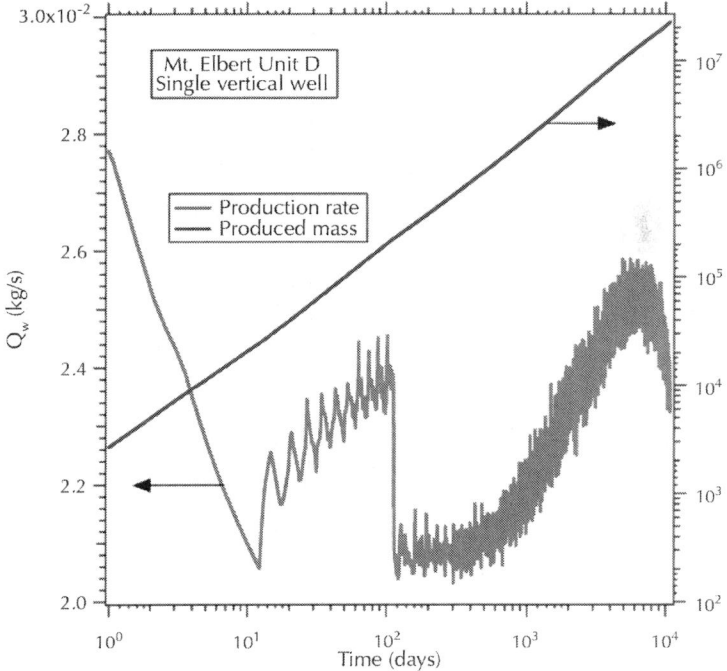

Figure. 4.3: Evolution of Q_W and M_W during production from unit D using a vertical well.

Of particular interest is the evolution of M_{HR} in Fig. 4.4, which indicates that barely 0.2% of the total hydrate mass in the HBL has dissociated at the end of the production period. In practical terms, this indicates that 30 years of continuous production have not even made a dent to the original hydrate mass. This is unequivocally demonstrated by the spatial distributions of the phase saturations shown in Fig. 4.5. The staircase appearance of the M_{HR} is the result of the very limited dissociation, to the point that discretization effects become evident: dissociation and hydrate depletion are characterized by very sharp fronts and occur in very few gridblocks, with their numbers too limited to result in a smoother curve.

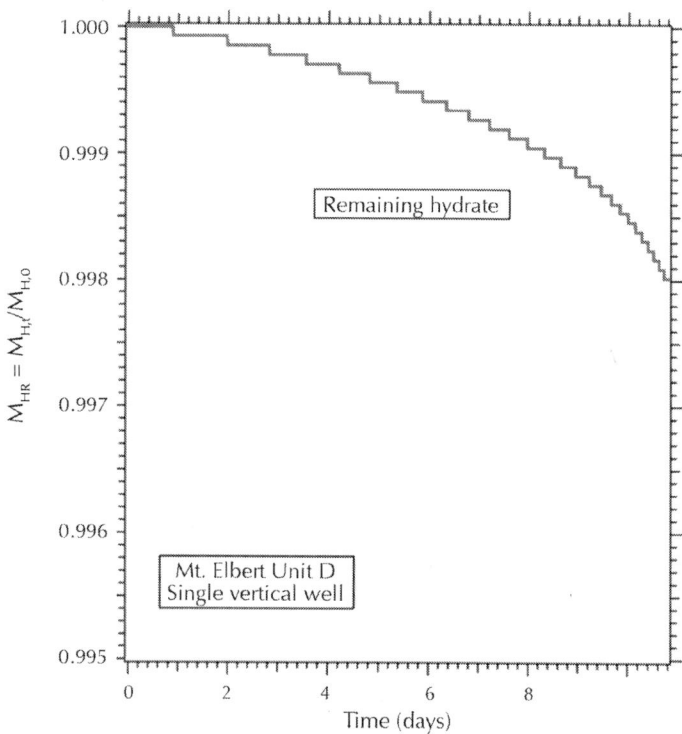

Figure. 4.4: Fraction of remaining hydrate mass M_{HR} vs. time during production from unit D using a vertical well. Note the negligible hydrate destruction after 30 years of continuous production.

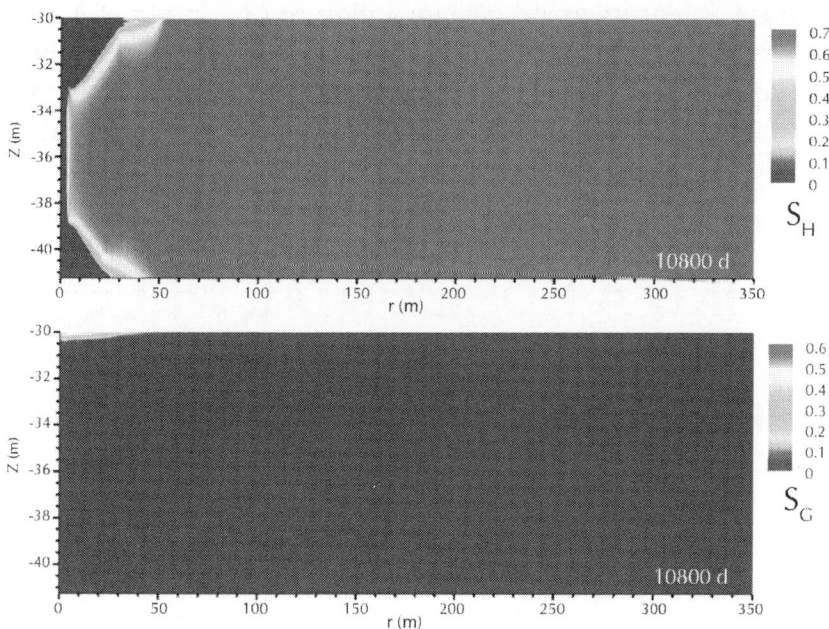

Figure. 4.5: S_H and S_G profiles in Unit D after 30 years of continuous production using a vertical well.

Phase Saturations and Overall Evaluation

The S_H distribution at $t = 30$ years (Fig. 4.5) depicts hydrate destruction that is minimal in extent and concentrated in the vicinity of the well at the top and bottom of the HBL, consistent with the very late onset of significant gas release in the reservoir seen in Fig. 4.1. The spatial distribution of S_G in Fig. 4.5 is consistent with the low V_F levels of Fig. 4.2. It is defined by (a) the accumulation of high-S_G gas in a limited zone at the top of the HBL, (b) very low S_G below the gas bank for $r < 50$ m, and (c) $S_G = 0$ in the rest of the profile. Drainage of water originating from dissociation and buoyancy of the released gas are the reasons for the absence of a zone of significant S_G at the bottom of the HBL near the well, even though the S_H profile shows evidence of dissociation. Clearly, gas production from such a relatively cold, permafrost-associated Class

3 deposit using vertical wells appears to be very ineffective, with little (if any) hope of attaining commercial viability. In the ensuing sections, we investigate the use of horizontal wells (operating at a constant P_w) as an alternative production strategy.

THE CASE OF PRODUCTION US-ING A HORIZONTAL WELL

Gas Production

Fig. 5.1 shows the evolution of Q_R and Q_p from a horizontal well (described in Fig. 3.2) over time, and includes for reference the Q_R and Q_p corresponding to the vertical well (from Fig. 4.1). The use of the horizontal well is shown to increase both Q_R and Q_p by about two orders of magnitude. While the improvement in performance over the vertical well is dramatic, Q_p remains low in absolute terms. However, it is possible that the production outlook may improve with longer wells, different well configurations, more complex production strategies, and by the consideration of heterogeneity (which has been shown to improve production in layered systems such as the ones in units C and D of Mount Elbert – see Anderson, 2011a; Kurihara et al., 2005 and Kurihara et al., 2009).

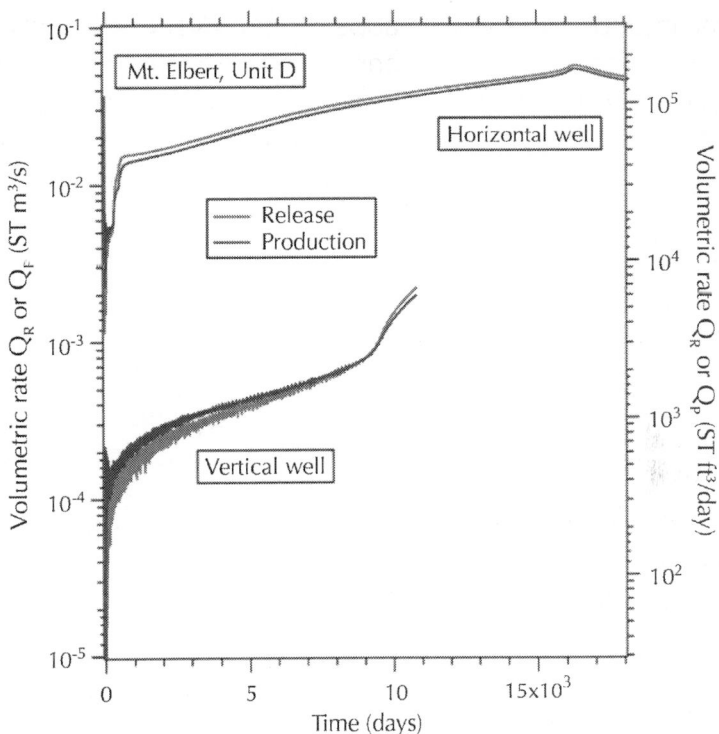

Figure. 5.1: Comparison of Q_R and Q_P from a vertical and a horizontal well during production from Unit D.

The evolution of Q_R and Q_P is characterized by an initial short period (Stage 1, to $t = 670$ days) of rapid increase, is succeeded by a long period of continuous but mild increase (stage 2) that lasts until $t = 16,300$ days (i.e., almost 45 years), and is followed by a period of continuous mild decline (Stage 3). At the end of Stage 1, $Q_P = 1.34 \times 10^{-2}$ ST m³/s (=4.1 × 10⁴ ST ft³/day), and peaks at the end of Stage 2, when $Q_P = 5.35 \times 10^{-2}$ ST m³/s (=1.63 × 10⁵ ST ft³/day).

Stage 1 is associated with rapid depressurization (especially near the wellbore) and corresponds to the rapid advancement of the depressurization front in the deposit (as will be shown in Section 5.3). Because (a) the pressure drop $P = P_0 - P_w$ between the bottomhole pressure and the pressure at the dissociation front is at its maximum

ΔP_{max} in the HBL, and (b) dissociation expands continuously into unaffected parts of the HBL as the depressurization front advances quickly, Q_R and Q_P increase rapidly and dQ_R/dt and dQ_P/dt are at their maximum. The endothermic nature of the hydrate dissociation reaction results in cooling of the HBL, but this has a limited effect in countering the effects of maximum ΔP on subsequent dissociation.

The end of Stage 1 and onset of Stage 2 is marked by the depressurization front reaching the outer boundaries of the HBL domain (i.e., at $y = L_y$, $x = L_x$). When this happens, the pressure wave can no longer advance, and the pressure drop at any point in the domain $\Delta P = P - P_w < \Delta P_{max}$. While Q_R and Q_P continue to increase because a larger volume of hydrate is dissociating, they do so slower, i.e., the lower pressure gradient leads to the reduction in dQ_R/dt and dQ_P/dt, which remain positive. Additionally, continuing HBL cooling caused by advancing hydrate dissociation makes further dissociation progressively more difficult.

Finally, the continuously diminishing driving force of dissociation (i.e., the ΔP) and the parallel reduction in the sensible heat that fuels and supports it lead to the declining Q_R and Q_P in Stage 3, which is characterized by negative dQ_R/dt and dQ_P/dt (Fig. 5.1). Because of the low T of the HBL in our study and, consequently, Q_R and Q_P are low in relation to the total hydrate mass, there is significant delay in the onset of Stage 3, and production increases monotonically and continuously over almost 45 years of production.

Q_R and Q_P in Fig. 5.1 are very similar in magnitude, as was the case in production from a vertical well. Unlike the vertical well case, $Q_R > Q_P$ in production from the horizontal well. We observe a similar pattern in the relationship of V_R and V_P in Fig. 5.2, with V_R being very slightly larger than V_P, while both (and V_F) are about two orders of magnitude larger than the ones corresponding to the vertical well case. Review of the relative magnitudes of Q_R, Q_P, V_R, V_P, and V_F confirms the pattern identified in the vertical well case, i.e., little gas accumulation in the reservoir, with most of the gas released from dissociation and dissolution is produced at the horizontal well. The cumulative produced volume V_P in Fig. 5.2 provides further confirmation of the improved outlook, but also of the challenge of unit D as a target for production from hydrates by depressurization: after continuous production for $t = 50$ years, V_P $= 5.3 \times 10^7$ ST m^3(=1.9 $\times$ 10^9 ST ft^3) of CH$_4$ have been produced. While this is a tremendous improvement over the vertical well case, it is still low in absolute terms.

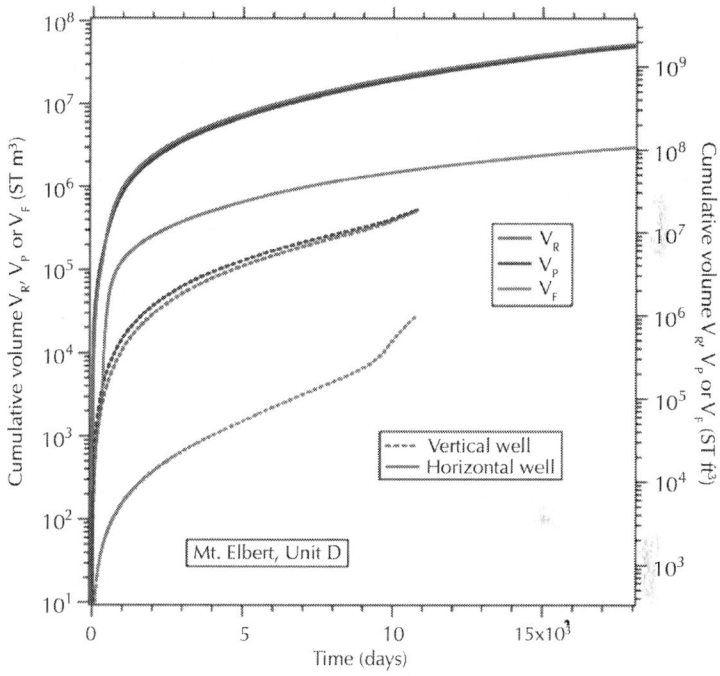

Figure. 5.2: Comparison of V_R, V_P and V_F from a vertical and a horizontal well producing from Unit D.

Water Production and Effectiveness of Dissociation

The water production rate Q_W in Fig. 5.3 shows some fluctuations at very early times ($T < 100$ days), but it then stabilizes at a low level and decreases slowly over the 50-year production period. During the entire time, Q_W is confined within a very narrow range

(0.8 kg/s < Q_W < 1.3 kg/s), and its declining long-term trend is (a) an inevitable consequence of a continuously declining pressure differential P, and (b) consistent with observations and conclusions from previous studies of production from hydrates (Moridis and Reagan, 2007a and Moridis and Reagan, 2007b). Of interest is the near parallel appearance of the cumulative water mass M_W curves in the horizontal and vertical cases. While M_W is larger (as expected) in the horizontal well case, it remains at manageable levels.

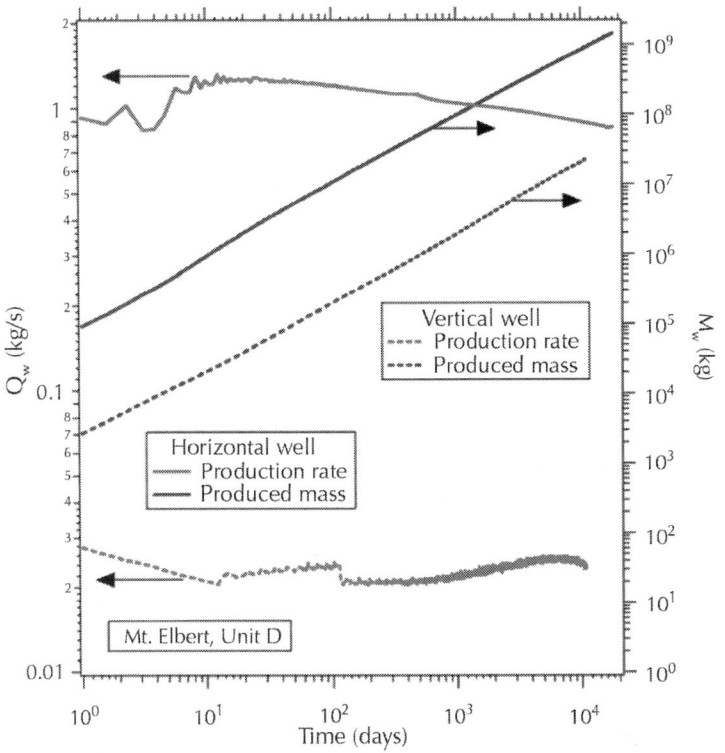

Figure. 5.3: Comparison of Q_W and M_W from a vertical and a horizontal well during production from Unit D.

Review of the evolution of M_{HR} in Fig. 5.4 provides additional evidence of the significant improvement in the effectiveness of dissociation using a horizontal well. Thus, about 9% of $M_{H,0}$ has been destroyed at t = 30 years, and the number rises to slightly over 20% after t = 50 years. While these numbers are respectable, they remain low when compared to production from conventional gas reservoirs employing horizontal wells of similar size. The inevitable conclusion is that, while horizontal wells are orders of magnitude more effective than vertical ones in gas production from hydrate deposits, relatively cold, permafrost-associated hydrates are challenging targets. Barring new developments in the technology of production from hydrates and changes in the pricing environment of natural gas, it is possible that such deposits may only hold promise as very-long-term, low- Q_p reservoirs.

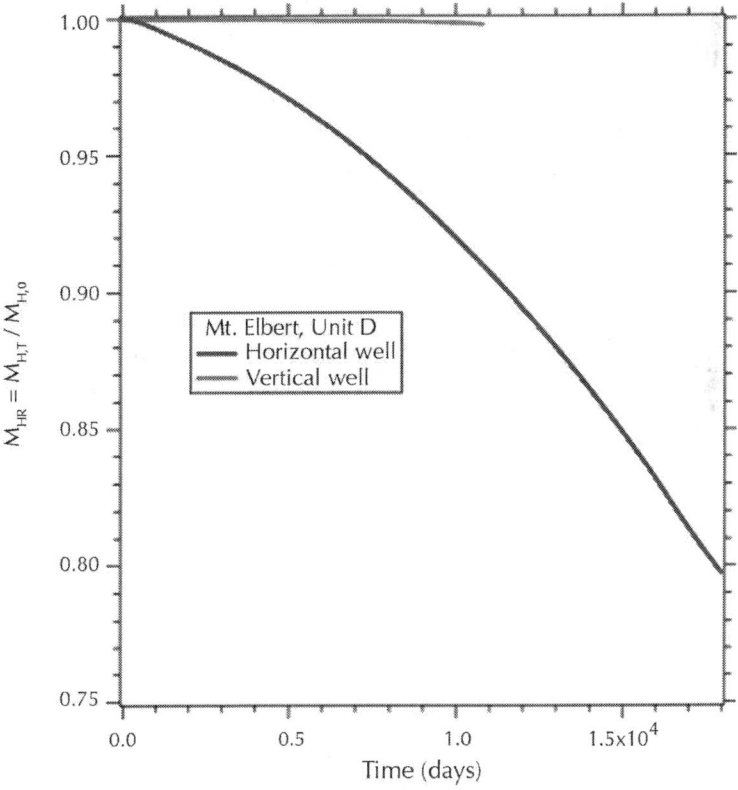

Figure. 5.4: Comparison of evolution of M_{HR} from a vertical and a horizontal well during production from Unit D.

Spatial Distributions

The evolution of the spatial distribution of S_H in Fig. 5.5 shows a dissociation pattern that is clearly far more extensive than that of the vertical well, but which is still limited compared to other studies involving much warmer deposits (Moridis and Reagan, 2007b). The minor dissociation observed until $t = 720$ days reflects the low T_0 of the unit D deposit. Note that significant hydrate dissociation occurs at the two main locations identified in the case of the vertical well: mainly at the HBL top, where the horizontal well is located (and where depressurization is at its maximum), and to a much lesser extent along the base of the HBL because of continuing geothermal heat inflows from the underburden.

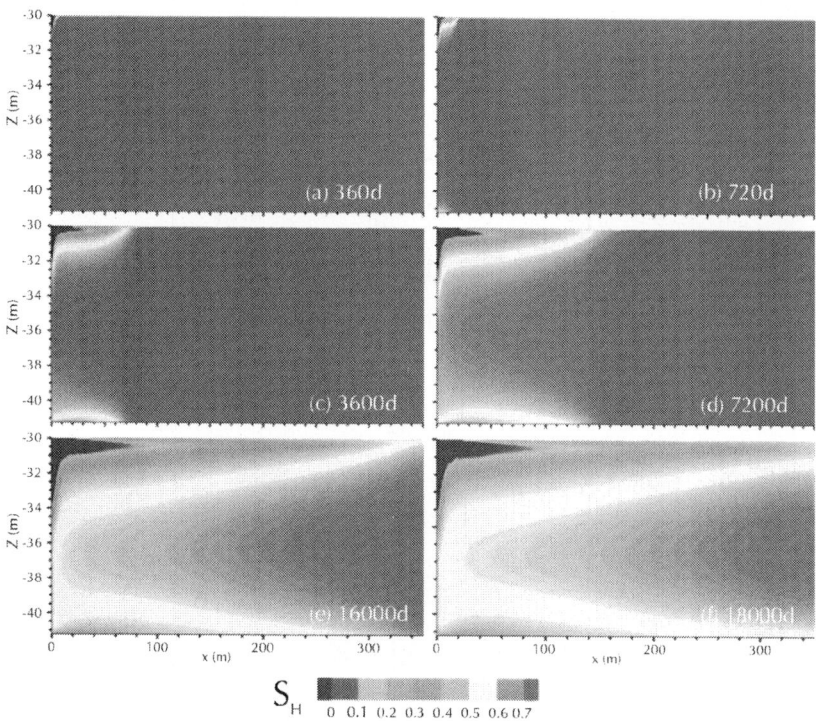

Figure. 5.5: Evolution of the spatial distribution of S_H in Unit D during 50 years of continuous production using a horizontal well.

The S_G distribution over time in Fig. 5.6 is consistent with the relatively low dissociation and limited gas accumulation indicated by Fig. 5.1 and Fig. 5.2, and depicted by Fig. 5.5. It takes a very long time ($t = 720$ days) for gas to accumulate at discernible saturation levels in the HBL. When this happens, it is limited to a thin gas zone at the top of the HBL. Unlike the case of the vertical well (in which the penetration of the entire HBL made possible the emergence of a free-gas zone at very low (almost trace) S_H levels at the base of the HBL, see Fig. 4.5), the significant physical separation of the horizontal well from the bottom of the HBL prevents the emergence of even traces of gas at this location because of buoyancy of the gas released from dissociation (rising to the top), and drainage of the water released along the entire HBL profile. A comparison of the S_H and S_G profiles in Fig. 5.5 and Fig. 5.6 indicates that the zone of significant hydrate destruction is much smaller than the free-gas zone because the water released from dissociation (rather than CH_4) occupies the newly hydrate-free space.

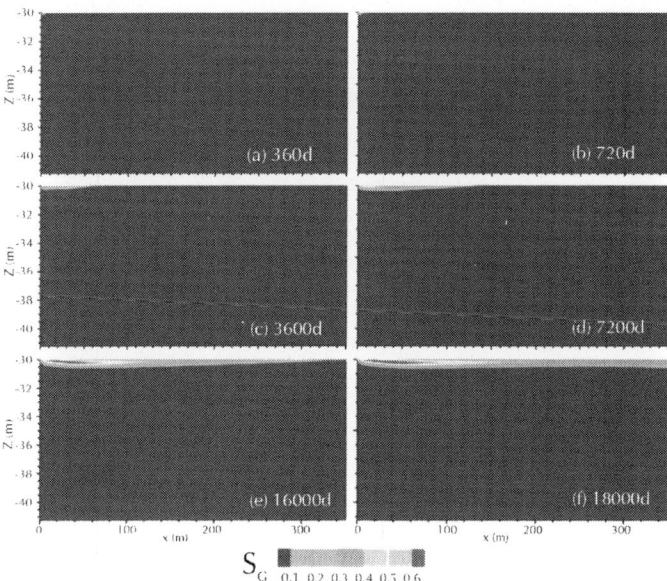

Figure. 5.6: Evolution of the spatial distribution of S_G in Unit D during 50 years of continuous production using a horizontal well.

The T distribution in Fig. 5.7 clearly describes the advancing dissociation interface as a sharp front that defines an abrupt temperature change within a narrow zone: while the T is practically undisturbed at near-T_0 levels ahead of the front until well into the production process (i.e., for $t > 7200$ days), the endothermic dissociation reaction leads to a significant temperature drop behind it, i.e., in the region where depressurization is causing dissociation. The sharpness of the interface is caused by the low effective permeability k_{eff} (caused by the high initial S_H) of the HBL. As expected, the sharp front (depicted by the abrupt temperature change in Fig. 5.7) moves away from the $x = 0$ axis in a manner that seems to correlate very well with the edges (top and bottom) of the dissociation zone depicted by the hydrate destruction in theS_H profile of Fig. 5.5. The sharp T front disappears sometime after $t = 7200$, when the dissociation front reaches the $x = L_x$ boundary and the entire domain begins to dissociate, albeit at a slow rate. The evolution of T over time in Fig. 5.7 is consistent with expectations, with the HBL becoming progressively colder as dissociation continues.

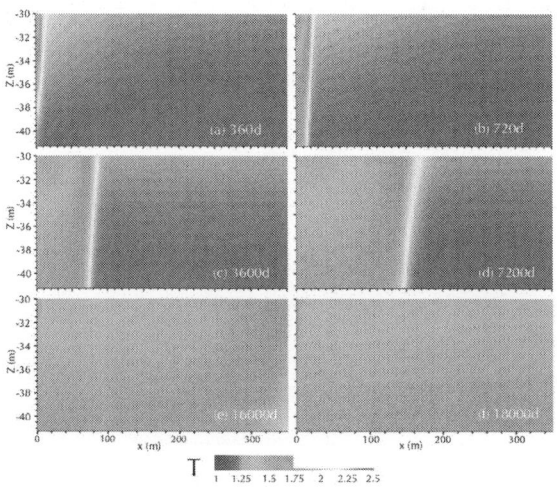

Figure. 5.7: Evolution of the spatial distribution of T in Unit D during 50 years of continuous production using a horizontal well.

The P distributions in Fig. 5.8 show that the entire HBL experiences a significant pressure drop even at early times, which, unlike the sharp fronts that mark the T distribution, occurs in a diffuse manner over extended regions. This is caused by the high-speed of propagation of the pressure wave in porous media, which is typical of advective processes of fluid flow. Comparison of the P distribution in Fig. 5.8 to the S_H and T distributions (in Fig. 5.5 and Fig. 5.7, respectively) shows that the pressure front advances well ahead of the dissociation front. This is caused by the initial thermodynamic state of the hydrate in the HBL, which is quite stable, i.e., safely within the Lw + H zone and away from the base of the stability zone defined by the Lw + H + V equilibrium curve of three-phase coexistence in Fig. 1.1. Because of the enhanced initial stability, a significant pressure drop has to be attained before dissociation can begin in earnest (and be marked by sharp T-fronts and disturbed S_H profiles), hence the lag in the initiation of dissociation.

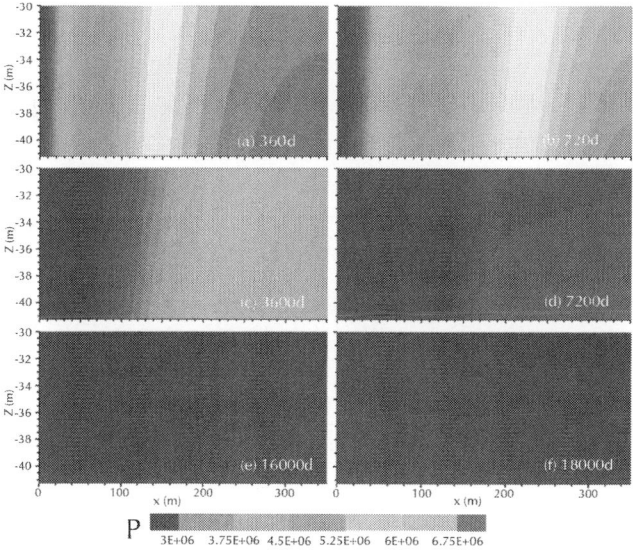

Figure. 5.8: Evolution of the spatial distribution of P in Unit D during 50 years of continuous production using a horizontal well.

SENSITIVITY ANALYSIS

Using the HBL and horizontal well described in Section 3 and analyzed in Section 5, we investigated the sensitivity of gas production to the following conditions and parameters:

- The stability of the hydrate deposit, as quantified by its initial temperature T and its deviation from the equilibrium temperature at the prevailing pressure,
- The initial hydrate saturation S_H,
- The formation anisotropy, i.e., the k_V/k_H ratio of the HBL sediment.

Because the first part of this study has clearly demonstrated the limited effectiveness of vertical wells, all sensitivity-related studies involved horizontal wells of the type shown in Fig. 3.2.

Sensitivity to *T*

Fig. 6.1 shows the dramatic effect that T (as a measure of the hydrate stability at a given P) has on production. For this study, the temperature of the HBL boundaries (T_T and T_B, see Section 3.5) of the D unit was raised by $\Delta T = 1\ °C$, resulting in a similar ΔT along the entire HBL profile. The increase in Q_R and Q_p corresponding to this slight temperature rise is nothing less than spectacular, reaching a factor of almost 8 at its peak. Because of the strong enhancing effect of the higher T on dissociation, Stage 1 and Stage 2 (associated with the Q_R and Q_p peak, see Section 5.1) occur much earlier than in the reference case, and, consequently, Stage 3 (which involves slow dissociation from the entire hydrate body in the HBL in response to a mild and declining ΔP) is longer and marked by a very gradual decline in Q_R and Q_p. Fig. 6.1 shows that the increase in Q_R and Q_p is most prevalent during the earlier part of the production period, i.e., during Stages 1 and 2.

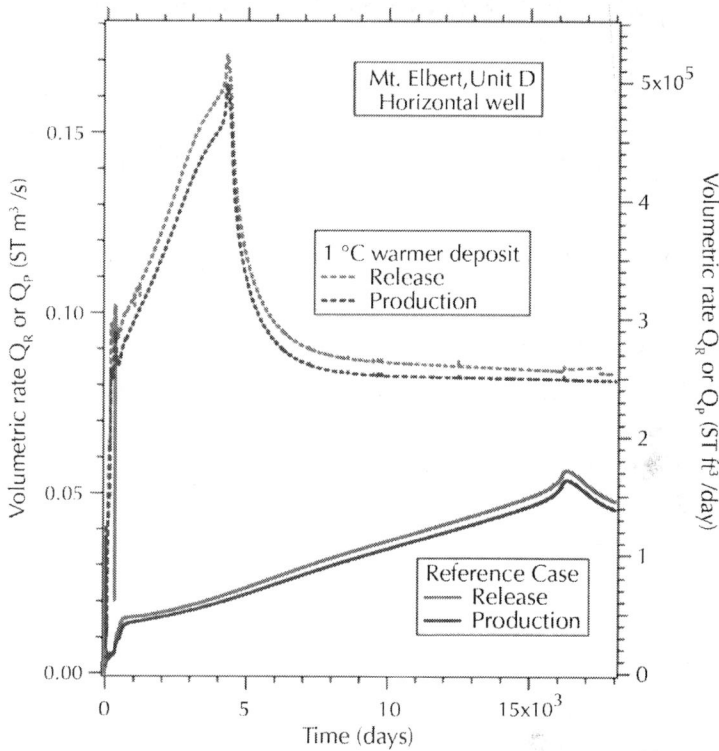

Figure. 6.1: Sensitivity of Q_R and Q_P to the initial temperature of the deposit.

The strong effect of even a slightly higher T on production is also demonstrated by Fig. 6.1, which shows that V_R and V_P increase (over the reference case) by a factor which is 2.8 at its minimum (at the end of the production period), and higher earlier in the production period. In essence, these results indicate the superiority of warmer hydrate deposits as potential production targets (in terms of production and early return), and are consistent with previous observations (Moridis and Reagan, 2007a, Moridis and Reagan, 2007b and Reagan et al., 2008). The increase in V_F is also significant in relative terms, but the total volume of free gas remains low (in absolute terms) and indicates that, as in the reference case, gas accumulation in the reservoir is limited because most of the released CH_4 is produced at the well.

The appeal of even slightly warmer deposits is further demonstrated by Fig. 6.3, which shows that water production remains practically the same despite drastically larger Q_R and Q_P. The M_{HR} in Fig. 6.4 shows that the $\Delta T = 1\ °C$ difference is sufficient to reduce the remaining hydrate from 80% to 40% of $M_{H,0}$, pointing to an increase in the mass of destroyed hydrate by a factor of 3 at the end of production. This is consistent with the results of Fig. 6.2, and encapsulates in a cumulative sense the appeal of warmer deposits. Further support of the superiority of warmer hydrates, in addition to visual confirmation of the results inFig. 5.2 and Fig. 5.4, is provided by the comparisons of the S_H and S_G profiles in Fig. 6.5 and Fig. 6.6, which show significant more hydrate destruction and larger free-gas accumulations for the case of the warmer HBL.

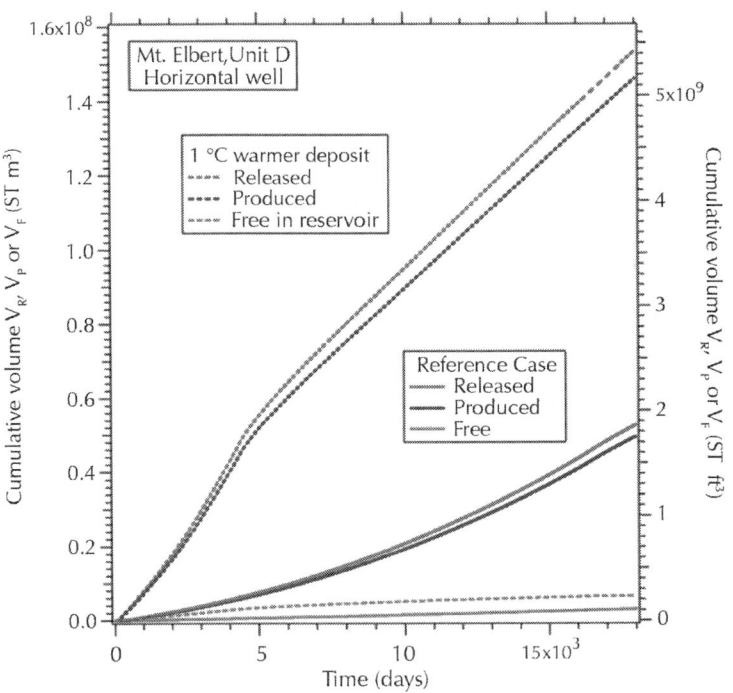

Figure. 6.2: Sensitivity of Q_R, Q_P and Q_F to the initial temperature of the deposit.

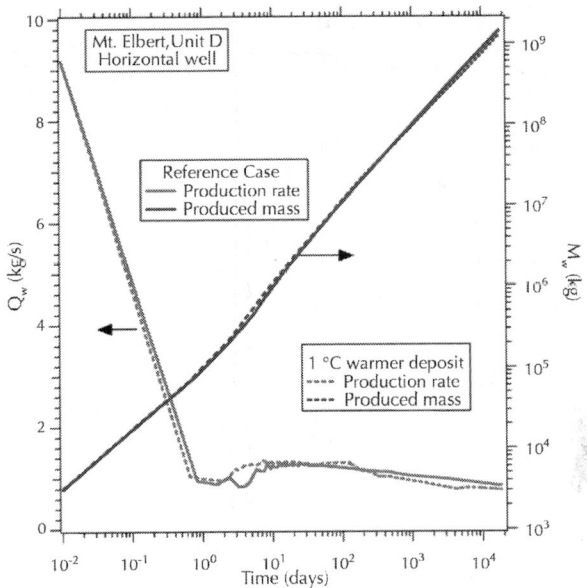

Figure. 6.3: Sensitivity of Q_W and M_W to the initial temperature of the deposit.

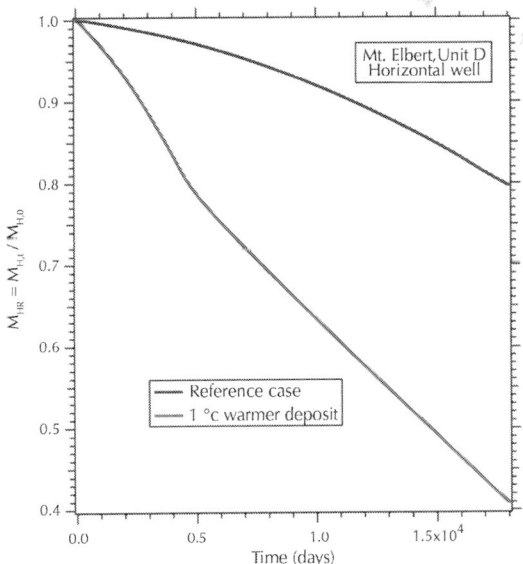

Figure. 6.4: Sensitivity of M_{HR} to the initial temperature of the deposit.

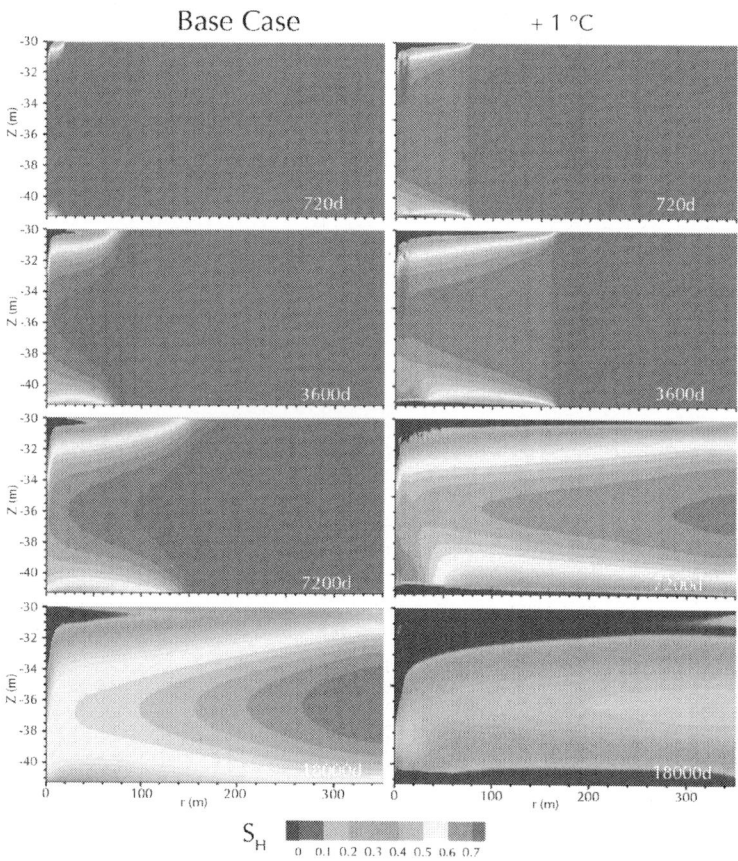

Figure. 6.5: Effect of the initial temperature on the evolution of the spatial distribution of S_H in Unit D during 50 years of continuous production using a horizontal well.

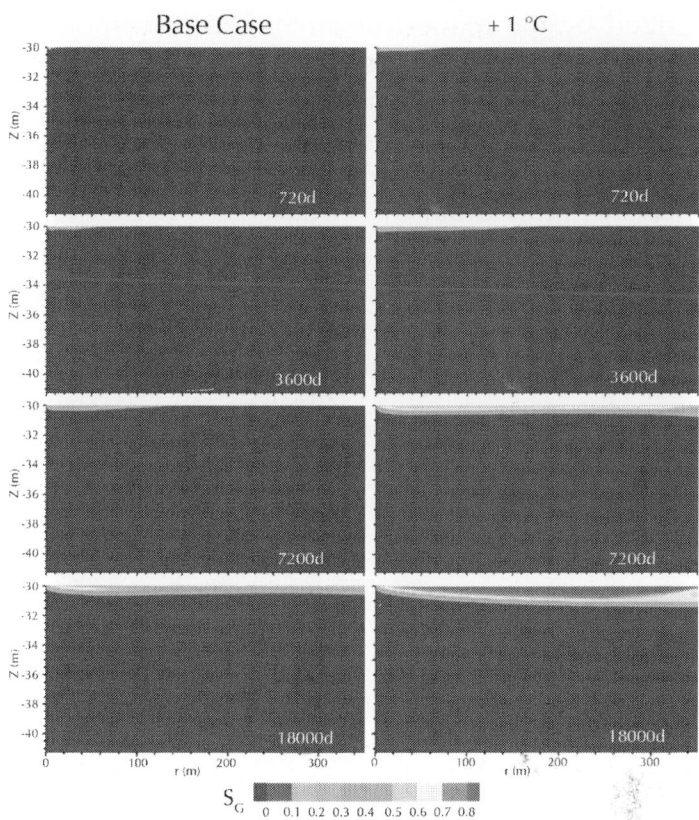

Figure. 6.6: Effect of the initial temperature on the evolution of the spatial distribution of S_G in Unit D during 50 years of continuous production using a horizontal well.

Sensitivity to S_H

Fig. 6.7 shows that Q_P increases with a decreasing S_H, at least within the range we investigated ($0.35 \leq S_H \leq 0.65$). This is attributed to the higher k_{eff} that corresponds to lower S_H levels for a given intrinsic permeability k. additionally, a decreasing S_H leads to correspondingly (and proportionally) shorter production Stages 1 and 2. Thus, the lowest $S_H = 0.35$ (a) has the highest Q_P that (b) occurs at the earliest time, and (c) has the longest Stage 3 (characterized by the mildest decline). The corresponding V_p vs. t curves in Fig. 6.8 show that the early high Q_p rates at low S_H are sufficient to preserve higher V_p despite later reversals in the relative Q_p magnitude, and provide further confirmation of the appeal of such "lean" hydrate systems under the conditions of the unit D Class 3 deposit at the Mount Elbert site.

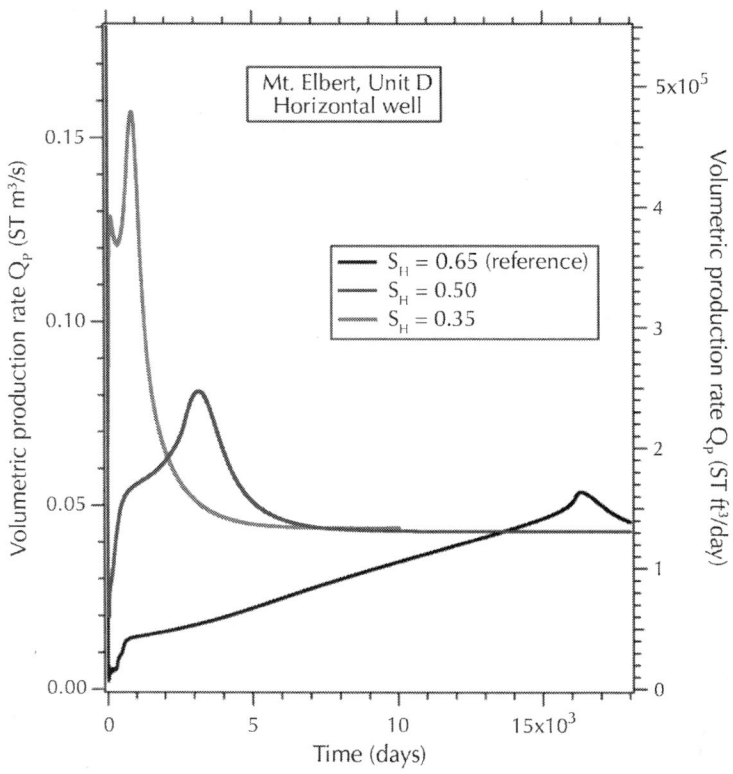

Figure. 6.7: Sensitivity of Q_p to the initial S_H of the deposit.

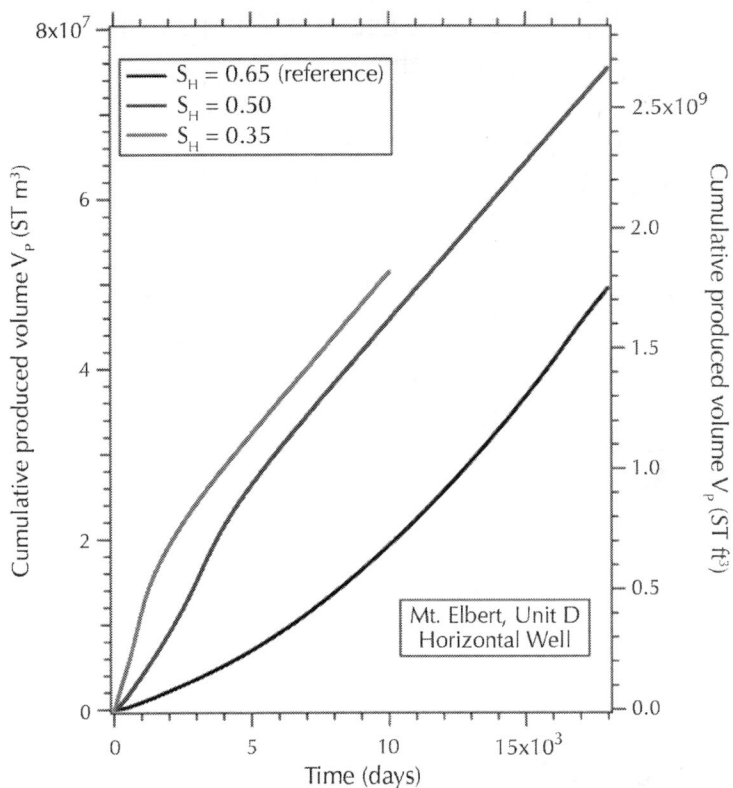

Figure. 6.8: Sensitivity of V_P to the initial S_H of the deposit.

The water production rates Q_w in Fig. 6.9 increase with a decreasing S_H. They differ by orders of magnitude for the various S_H in the study because they reflect drastically different k_{eff} regimes (strongly influenced by S_H). Q_w decreases over time because the driving force ΔP in the reservoir decreases as depressurization advances, and eventually the three Q_w appear to converge after $t = 10,000$ days of production. Similarly, the cumulative mass of produced water M_w increases with a decreasing S_H, but eventually the three M_w curves converge at about $t = 5000$ days. The M_{HR} curve in Fig. 6.10 provides a measure of the relative advantage that lower S_H confer to gas production from hydrates with the attributes if unit D, and is complementary to the results in Fig. 6.8.

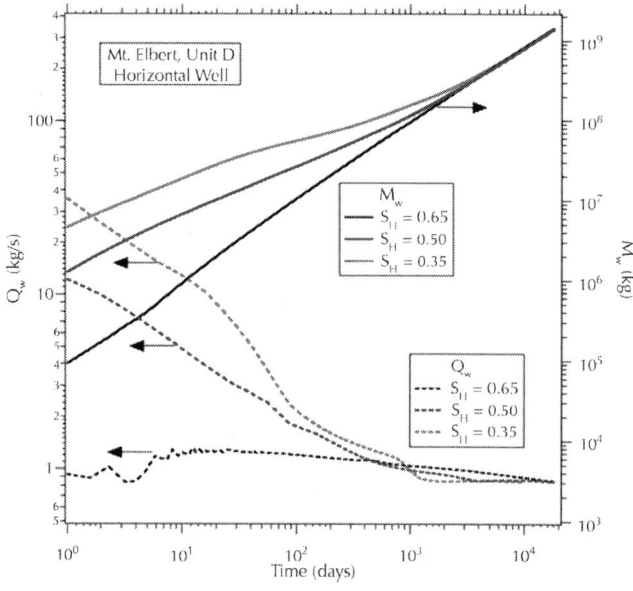

Figure. 6.9: Sensitivity of Q_W and M_W to the initial S_H of the deposit.

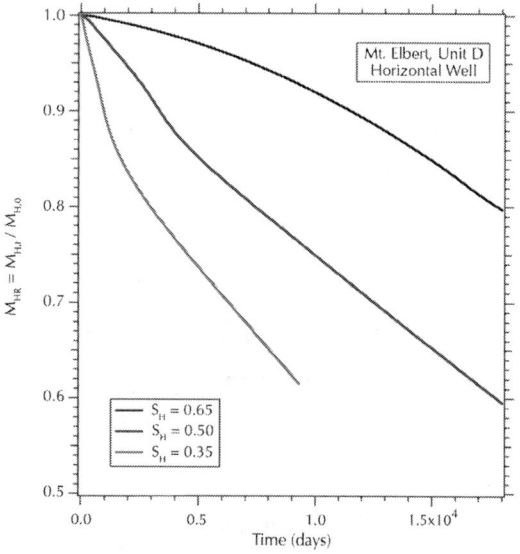

Figure. 6.10: Sensitivity of M_{HR} to the initial S_H of the deposit.

Sensitivity to Anisotropy

Review of the evolution of the S_H distribution pattern in Fig. 5.5 showed the propensity of dissociation to advance horizontally and preferentially along the top (mainly) and the bottom of the HBL because of the heat inflows from the boundaries. At the top of the HBL, this tendency was significantly enhanced by the proximity to the well, where depressurization is at its most intense, and by the favorable relative permeability to gas flow (a result of gas accumulation at this location). Given these earlier indications, the results of the effect of anisotropy (described by the ratio $k_R = k_V/k_H$) in Fig. 6.11 are entirely anticipated. Decreasing k_R to 0.1 (from its original value of 1 in the reference case) creates a flow regime that enhances horizontal flow and facilitates dissociation.

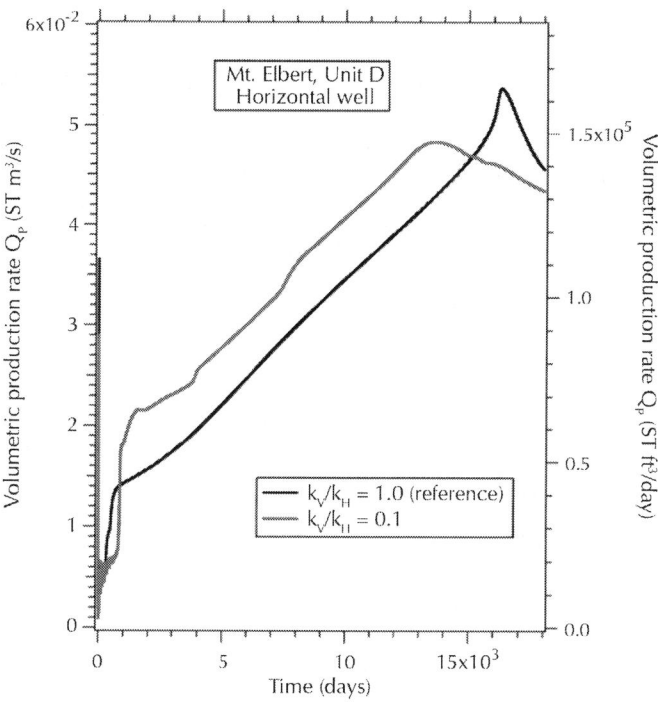

Figure. 6.11: Sensitivity of Q_P to $k_R = k_V/k_H$.

Thus, Q_p during Stage 2 (the longest of the 3 production stages) is higher than that in the reference case because dissociation and horizontal flow are favored. However, because the initial flow regime around the well immediately after the initiation of production is cylindrical, Q_p at a very early stage is lower than that in the reference case, and the lower cumulative permeability delays the time of arrival of the depressurization front at the $x = L_x$ boundary and prolongs Stage 1. However, the transition into the favorable horizontal dissociation and flow regime results in a Q_p that is higher than that for $k_R = 1$ in the later part of Stage 1. The more effective dissociation for $k_R = 0.1$ leads to faster cooling, resulting in an earlier onset of Stage 3 and a lower Q_p than in the reference case. The effect of k_R on production is depicted by the comparison of the corresponding V_F curves in Fig. 6.12. At early times ($t < 1400$ days), V_F for the $k_R = 1$ case exceeds that for $k_R = 0.1$. Beyond that time, the pattern reverses, and the lower k_R appears to have an advantage in long-term production. However, the increase in long-term production caused by the lower k_R is incremental and not that significant. This is further confirmed by the evolution of M_{HR} in Fig. 6.13, which shows only minor incremental hydrate destruction for $k_R = 0.1$.

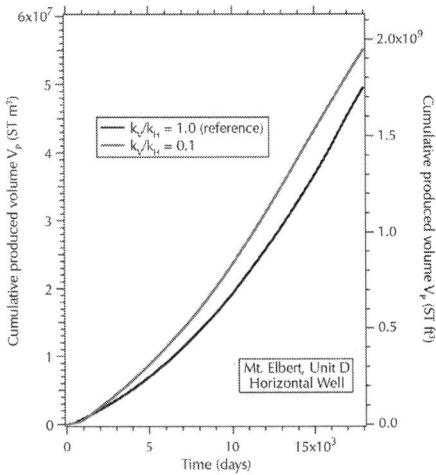

Figure. 6.12: Sensitivity of V_p to $k_R = k_V/k_H$.

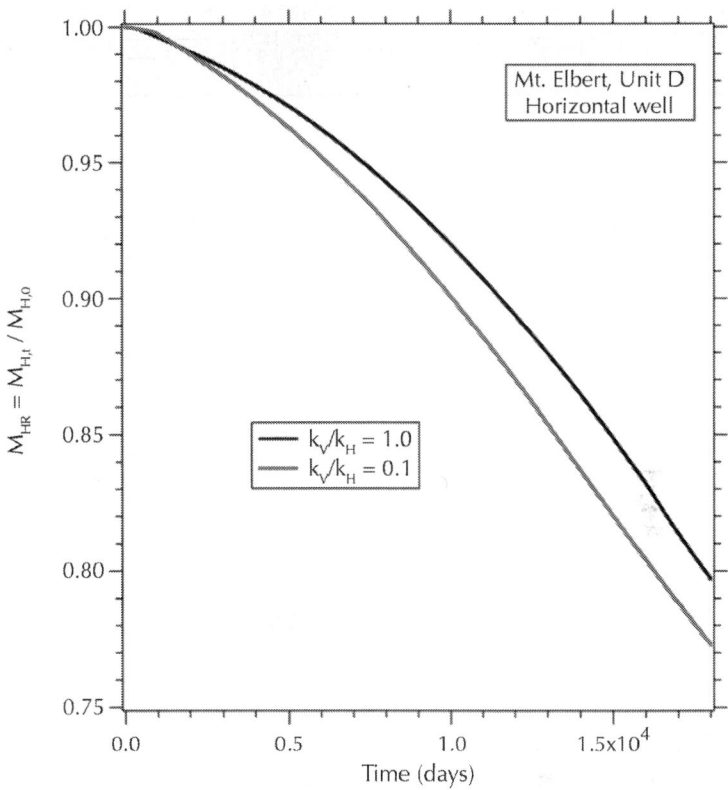

Figure. 6.13: Sensitivity of M_{HR} to $k_R = k_V/k_H$.

The water production rates Q_w for the two k_R levels in Fig. 6.14 follow the same pattern of long-term decline (a direct consequence of the declining ΔP differential), and their comparisons indicates that, all else being equal, Q_w decreases with a decreasing k_R. This is caused by the reduced permeability, which prevents the aqueous phase (widely distributed in the entire HBL, unlike the gas phase that is concentrated at the top – see Fig. 5.6) from flowing easily to the well. The M_w patterns in Fig. 6.14 reflect the Q_w relationship.

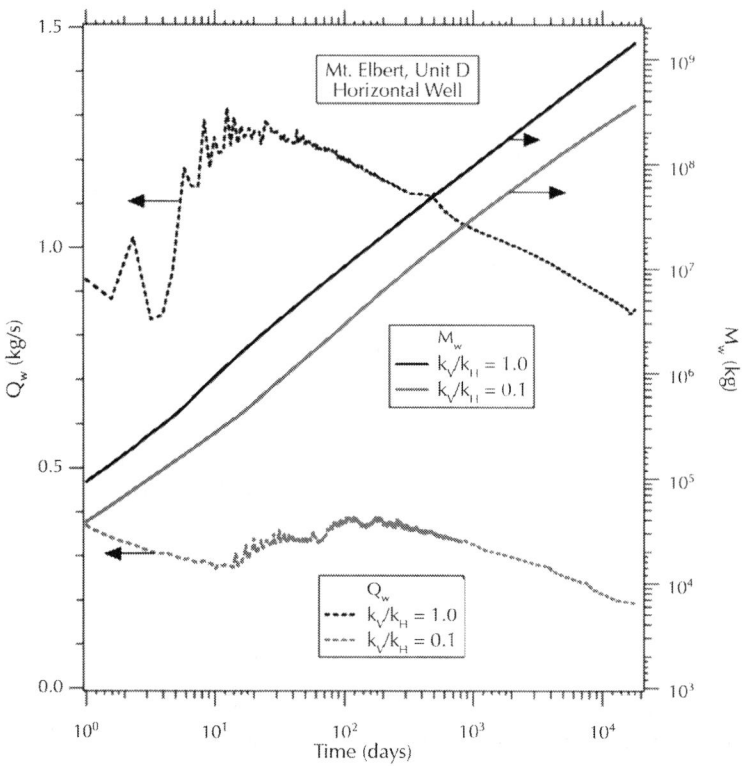

Figure. 6.14: Sensitivity of Q_w and M_w to $k_R = k_V/k_H$.

SUMMARY AND CONCLUSIONS

This study is part of an effort led by BPXA and the U.S. Department of Energy to identify appropriate targets for a long-term field test of production from permafrost-associated hydrate deposits. We focus on the evaluation of the gas production potential of the unit D hydrate accumulation at the Mount Elbert site, North Slope, Alaska, a shallower and colder Class 3 deposit than the unit C deposit. We investigate the performance of vertical and horizontal wells operating under constant bottomhole pressure in gas production fueled by depressurization-induced dissociation of the hydrates. Based on the results of this study, we draw the following conclusions:

- The effectiveness of vertical wells operating at a constant P_w in the low-P, low-T unit D deposit is very limited. Gas production from this hydrate accumulation is hampered by very low rates that persist for very long times

- Horizontal wells operating at a constant P_w appear to yield higher production rates relative to vertical wells in the cold hydrate deposit of unit D at the Mount Elbert site. Although production using horizontal wells is about two orders of magnitude larger than that from vertical wells accessing the same section of the HBL, it is still low in absolute terms, and carries the additional burden of the significantly higher costs of installing and operating a horizontal well.

- Water production in either the vertical or the horizontal well case remains well within manageable limits.

- It is possible that the production rates may improve with the use of longer wells, the use of different well configurations, and the development of more complex strategies to deliver more efficient dissociation. Such new developments and solutions are not evident. Although the authors are aware of some new ideas that have been proposed to address these issues, their technical feasibility has not been explored (let alone established), and their potential effectiveness in increasing gas production has yet to be evaluated.

- The sensitivity analysis we conducted identified the desirable features, to be used as criteria for the selection of a hydrate deposit as an appropriate production target. The sensitivity of gas production to the initial HBL temperature is nothing short of dramatic: a $\Delta T = 1$ °C increase results in a 8-fold increase in Q_p, and in a 3-fold (at least) increase in the V_p, while water production is practically unaffected. Confirming earlier studies (Moridis and Reagan, 2007a, Moridis and Reagan, 2007b and Moridis et al., 2008b), temperature emerges again as the most important factor affecting gas production and its importance appears enhanced in the case of colder deposits. These results and observations guide us to select the deepest, warmest hydrate deposit (i.e., with the highest sensible heat and affording the largest possible pressure gradient ΔP) as the most promising production target from among those accumulations that meet other basic criteria of reservoir quality and accessibility.

- All other things being equal, hydrate accumulations with low S_H appear to be more desirable potential targets for a successful long-term field test of production from colder, permafrost-associated hydrates because of their tendency to yield higher Q_P at early times that are well within the time frame of the planned test. Water production increases with a decreasing S_H, but converges to the same level in the long run.

- Anisotropy is not a feature that is readily observable, and is difficult to use as a criterion for the selection of an appropriate hydrate deposits as a production target. If there is evidence of anisotropy caused by external factors (such as sedimentation patterns, which usually result in $k_R < 1$), this is expected to lead to higher gas production (and lower water production) within the time frame of the planned field test.

ACKNOWLEDGMENT

This work was supported by the Assistant Secretary for Fossil Energy, Office of Natural Gas and Petroleum Technology, through the National Energy Technology Laboratory, under the U.S. Department of Energy, Contract No. DE-AC02-05CH11231. The authors are indebted to John Apps and Dan Hawkes for their careful review.

REFERENCES

1.   Anderson, B.J., Hancock, S.H., Wilson, S.J., Enger, C.S., Collett, T.S., Boswell, R.M., Hunter, R.B., 2011a. Formation pressure testing at the Mount Elbert Gas Hydrate Stratigraphic Test Well, Alaska North Slope: Operational summary,

history matching, and interpretations. Journal of Marine and Petroleum Geology 28 (2), 478–492.

2. Anderson, B., Kurihara, M., White, M.D., Moridis, G.J., Wilson, S.J., Pooladi-Darvish, M., Gaddipati, M., Masuda, Y., Collett, T.S., Hunter, R.B., Narita, H., Rose, K., Boswell, R., 2011b. Regional long-term production modeling from a single well test, Mount Elbert Gas Hydrate Stratigraphic Test Well, Alaska North Slope. Marine and Petroleum Geology 28 (2), 493–501.

3. Anderson, B.J., Wilder, J.W., Kurihara, M., White, M.D., Moridis, G.J., Wilson, S.J., Pooladi Darvish, M., Masuda, Y., Collett, T.S., Hunter, R.B., Narita, H., Rose, K., Boswell, R., 2008. Analysis of modular dynamic formation test results from the Mount Elbert-01 stratigraphic test well, Milne Point Unit, North Slope, Alaska. In: Paper Presented at the 6th International Conference on Gas Hydrates, Vancouver, British Columbia, Canada, July 6–10, 2008. Bird, K.J., Magoon, L.B., 1987. Petroleum Geology of the Northern Part of the Arctic National Wildlife Refuge, vol. 1778. U.S. Geological Survey Bulletin, Northeastern Alaska, pp. 324.

4. Boswell, R., Hunter, R., Collett, T.S., Digert, S., Hancock, S., Weeks, M., Mount Elbert Science Team, 2008. Investigation of gas hydrate bearing sandstone reservoirs at the Mount Elbert stratigraphic test well, Milne Point, Alaska. In: Proceedings of the 6th International Conference on Gas Hydrates, July 6–10, 2008, Vancouver, British Columbia, Canada.

5. Boswell, R.M., Rose, K.K., Collett, T.S., Lee, M.W., Winters, W.J., Lewis, K.A., Agena, W.F., 2011. Geologic controls on gas hydrate occurrence in the Mount Elbert prospect, Alaska North Slope. Journal of Marine and Petroleum Geology 28 (2), 589–607.

6. Clarke, M.A., Bishnoi, P.R., 2000. Determination of the intrinsic rate of methane gas hydrate decomposition. Chemical Engineering Science 55, 4869.

7. Collett, T., 1993. Natural gas hydrates of the Prudhoe Bay

and Kuparuk River area, North Slope, Alaska. American Association of Petroleum Geologists Bulletin 77 (5), 793–812.

8.   Collett, T., 2007. Arctic gas hydrate energy assessment studies. In: The Arctic Energy Summit, Anchorage, Alaska, 15–18 October 2007.

9.   Collett, T.S., Lee, M.W., Agena, W.F., Miller, J.J., Lewis, K.A., Zyrianova, M.V., Boswell, R.M., Inks, T.L., 2011a. Permafrost-associated natural gas hydrate occurrences on the Alaska North Slope. Journal of Marine and Petroleum Geology 28 (2), 279–294.

10.   Collett, T.S., Lewis, R.E., Winters, W.J., Lee, M.W., Rose, K.K., Boswell, R.M., 2011b. Downhole well log and core montages from the Mount Elbert Gas Hydrate Stratigraphic Test Well, Alaska North Slope. Journal of Marine and Petroleum Geology 28 (2), 561–577.

11.   Collett, T.S., Bird, K.J., Kvenvolden, K.A., Magoon, L.B., 1988. Geologic interrelations relative to gas hydrates within the north slope of Alaska: U.S. Geological Survey Open-File Report 88–389, 150.

12.   Dallimore, S.R., Collett, T.S. (Eds.), 2005. Scientific Results from the Mallik 2002 Gas Hydrate Production Research Well Program, Mackenzie Delta, Northwest Territories, Canada. Geological Survey of Canada Bulletin, vol. 585.

13.   Dallimore, S.R., Uchida, T., Collett, T.S., 1999. Scientific results from JAPEX/JNOC/GSC Mallik 2L-38 gas hydrate research well, Mackenzie delta, northwest territories, Canada. Geological Survey of Canada Bulletin 544, 403.

14.   Hong, H., Pooladi-Darvish, M., 2005. Simulation of depressurization for gas production from gas hydrate reservoirs. Journal of Canadian Petroleum Technology 44 (11), 39–46.

15.   Hunter, R.B., Collett, T.S., Boswell, R.M., Anderson, B.J., Digert, S.A., Pospisil, G., Baker, R.C., Weeks, L.M., 2011. Mount Elbert Gas Hydrate Stratigraphic Test Well, Alaska North Slope: Overview of scientific and technical program.

Journal of Marine and Petroleum Geology 28 (2), 295–310.

16. Kim, H.C., Bishnoi, P.R., Heidemann, R.A., Rizvi, S.S.H., 1987. Kinetics of methane hydrate decomposition. Chemical Engineering Science 42 (7), 1645.

17. Klauda, J.B., Sandler, S.I., 2005. Global distribution of methane hydrate in ocean sediment. Energy and Fuels 19, 469.

18. Kowalsky, M.B., Moridis, G.J., 2007. Comparison of kinetic and equilibrium reactions in simulating the behavior of gas hydrates. Energy Conversion and Management 48, 1850. doi:10.1016/j.enconman.2007.01.017 (LBNL-63357).

19. Kurihara, M., Funatsu, K., Ouchi, H., Masuda Y., Narita, H., 2005. Investigation on applicability of methane hydrate production methods to reservoirs with diverse characteristics. In: Paper 3003 presented at the 5th International Conference on Gas Hydrates, Trondheim, Norway, 13–16 June, Proceedings, vol. 3, pp. 714–725.

20. Kurihara, M., Sato, A., Ouchi, H., Narita, H., Masuda, Y., Saeki, T., Fujii, T. 2008.

21. Prediction of Gas Productivity from Eastern Nankai Trough Methane-Hydrate

22. Reservoirs. Paper OTC 19382 presented at the Offshore Technology Conference,

23. Houston, 5-8 May. doi:10.4043/19382-MS.

24. Kurihara, M., Sato, A., Ouchi, H., Narita, H., Masuda, Y., Saeki, T., Fujii, T., 2009. Prediction of gas productivity from eastern Nankai trough methane-hydrate reservoirs. SPE Reservoir Evaluation and Engineering 12 (3), 477–499. doi:10.2118/125481-PA. SPE-125481-PA.

25. Makogon, Y.F., 1987. Gas hydrates: frozen energy. Recherche 18 (192), 1192.

26. Makogon, Y.F., 1997. Hydrates of Hydrocarbons. Penn Well Publishing Co, Tulsa, OK.

27. Milkov, A.V., 2004. Global estimates of hydrate-bound gas

in marine sediments: how much is really out there? Earth Science Reviews 66 (3), 183.

28. Moridis, G.J., 2003. Numerical studies of gas production from methane hydrates. SPE Journal 32 (8), 359.

29. Moridis, G.J., Collett, T., Dallimore, S., Satoh, T., Hancock, S., Weatherhill, B., 2004.

30. Numerical studies of gas production from several methane hydrate zones at the

31. Mallik Site, Mackenzie Delta, Canada. JPSE 43, 219.

32. Moridis, G.J., Collett, T.S., Boswell, R., Kurihara, M., Reagan, M.T., Koh, C., Sloan, E.D., 2008a. Toward production from gas hydrates: status, technology, and potential, SPE 114163. In: SPE Unconventional Reservoirs Conference, Keystone, Colorado, U.S.A., 10–12 February 2008.

33. Moridis, G.J., Kowalsky, M., Pruess, K., 2008b. Depressurization-induced gas production from Class 1 hydrate deposits. SPE Reservoir Evaluation and Engineering 10 (5), 458–488.

34. Moridis, G.J., Kowalsky, M.B., Pruess, K., 2008c. TOUGH þ HYDRATE v1.0 User's

35. Manual: A Code for the Simulation of System Behavior in Hydrate-Bearing Geologic Media, Report LBNL-00149E. Lawrence Berkeley National Laboratory, Berkeley, CA.

36. Moridis, G.J., Reagan, M.T., 2007a. Strategies for gas production from oceanic Class 3 hydrate accumulations, OTC-18865. In: 2007 Offshore Technology Conference,

37. Houston, Texas, 30 April–3 May 2007.

38. Moridis, G.J., Reagan, M.T., 2007b. Gas production from oceanic Class 2 hydrate accumulations, OTC 18866. In: 2007 Offshore Technology Conference, Houston, Texas, U.S.A., 30 April–3 May 2007.

39. Moridis, G.J., Reagan, M.T., 2007c. Gas production from Class 2 hydrate accumulations in the permafrost, SPE 110858. In: 2007 SPE Annual Technical Conference and Exhibition,

Anaheim, California, U.S.A., 11–14 November 2007.

40. Moridis, G.J., Sloan, E.D., 2007. Gas production potential of disperse low-saturation hydrate accumulations in oceanic sediments. Energy Conversion and Management 48 (6), 1834–1849.

41. Reagan, M.T., Moridis, G.J., Zhang, K., 2008. Sensitivity analysis of gas production from Class 2 and Class 3 hydrate deposits, OTC 19554. In: 2008 Offshore Technology Conference, Houston, Texas, USA, 5–8 May 2008. van Genuchten, M.Th, 1980. A closed-form equation for predicting the hydraulic conductivity of unsaturated soils. Soil Science Society of America 44, 892.

42. Sloan, E.D., Koh, C., 2008. Clathrate Hydrates of Nautral Gases, third ed. Taylor and

43. Francis, Inc., Boca Raton, FL.

44. Sun, X., Mohanty, K.K., 2005. Simulation of methane hydrate reservoirs, SPE 93015. In: 2005 SPE Reservoir Simulation Symposium, Houston, TX U.S.A., 31 January– 2 February 2005.

45. White, M., 2008. Personal communication.

46. Winters, W., Walker, M., Hunter, R., Collett, T.S., Boswell, R., Rose, K., Waite, W.,

47. Torres, M., Patil, S., Dandekar, A., 2011. Physical properties of sediment from the

48. BPXA-DOE-USGS Mount Elbert gas-hydrate stratigraphic test well. Marine and

49. Petroleum Geology 28 (2), 361–380.

50. Wright, J.F., Dallimore, S.R., Nixon, F.M., 1999. Influences of grain size and salinity on pressure-temperature thresholds for methane hydrate stability in JAPEX/JNOC/GSC Mallik 2L-38 gas hydrate research-well sediments. In: Dallimore, S.R., Uchida, T., Collett, T.S. (Eds.), Scientific Results from JAPEX/ JNOC/GSC Mallik 2L-38 Gas Hydrate Research-well, Mackenzie Delta, Northwest Territories, Canada. Geological Survey of Canada Bulletin, vol. 544, p. 229.

51. Zhang, K., Moridis, G.J., 2008. A domain decomposition approach for large-scale simulations of coupled processes in hydrate-bearing geologic media. In: Paper Presented at the 6th International Conference on Gas Hydrates, Vancouver, British Columbia, Canada, July 6–10, 2008.

Simulation of Gas Transport in Tight/Shale Gas Reservoirs by a Multicomponent Model Based on PEBI Grid

Longjun Zhang[1], Daolun Li[1, 2], Lei Wang[3], and Detang Lu[1]

[1]Department of Modern Mechanics, University of Science and Technology of China, Hefei 230027, China
[2]Hefei University of Technology, Hefei 230026, China
[3]Institute of Nuclear Energy Safety Technology, Chinese Academy of Sciences, Hefei 230031, China

ABSTRACT

The ultra-low permeability and nanosize pores of tight/shale gas reservoir would lead to non-Darcy flow including slip flow, transition flow, and free molecular flow, which cannot be described by traditional Darcy's law. The organic content often adsorbs some gas content, while the adsorbed amount for different gas species is different. Based on these facts, we develop a new compositional model based on unstructured PEBI (perpendicular bisection) grid, which is able to characterize non-Darcy flow including slip flow, transition flow, and free molecular flow and the multicomponent adsorption in tight/shale gas reservoirs. With the proposed model, we study the effect of non-Darcy flow, length of the hydraulic fracture, and initial gas composition on gas production. The results show both non-Darcy flow and fracture length have significant influence on gas production. Ignoring non-Darcy flow would underestimate 67% cumulative gas production in lower permeable gas reservoirs. Gas production increases with fracture length. In lower permeable reservoirs, gas production increases almost linearly with the hydraulic fracture length. However, in higher permeable reservoirs, the increment of the former gradually decreases with the increase in the latter. The results also show that the presence of CO_2 in the formation would lower down gas production.

INTRODUCTION

Gas production from unconventional gas reservoirs, such as tight gas/shale gas reservoir, has grown great interest in recent years. Because of the ultra-low permeability (usually under 0.1 mD) and small pore diameter (usually under 50 nm) [1], gas flow in such tight formations reveals multiflow mechanisms that cannot be described by traditional Darcy's law, such as slip flow and Knudsen diffusion [2, 3]. Some modeling work has been conducted to study flow mechanisms in such reservoirs. Javadpour [2] combined convective flow and Knudsen diffusion into gas mass balance equation and found that the apparent permeability derived from the new mass balance equation can lead to one to two orders of magnitude difference from the intrinsic permeability in origin Darcy's law. Beskok and Karniadakis [4] derived a unified Hagen-Poiseuille-type equation for volumetric gas flow through a single pipe. Based on Beskok and Karniadakis [4], Florence et al. [5] proposed a formulation of apparent permeability in terms of Knudsen number. Civan [6] improved the function of the dimensionless rarefaction coefficient proposed by Beskok and Karniadakis [4] and established a mathematical model for gas flow in tight gas formation [7]. Zheng et al. [8–10] proposed a predictive model for gas slippage factor and gas diffusivity in microporous media based on fractal theory. Freeman et al. [11] incorporated the dusty-gas model into TOUGH+ family code to study gas flow behavior in tight gas/shale gas reservoirs. Freeman et al. [12] also incorporate extended Langmuir isotherm into a compositional model to represent gas desorption in shale. However, they did not consider multiflow mechanisms this time, such as slip flow and transition flow. Clarkson et al. [13] modeled transport in tight gas/shale using dynamic slippage concept which developed by Ertekin et al. [14]. Swami et al. [15] and Li et al. [16] both separately incorporated multiflow mechanisms into a numerical model to simulate gas behavior in shale. Yao et al. [17] compared gas production predicted by Civan [6], Javadpour [2], and Dusty-gas model and studied effect of fracture parameters on gas production in shale. However, most models above are single component model and are based on structured grid [2, 7, 13,15–17]. Some models [2, 11, 12] did not combine multiflow mechanisms and gas sorption together. In this paper, we first developed a compositional model which incorporates multiflow mechanisms and multicomponent adsorption based on PEBI (perpendicular bisection) grid. We also studied effect of apparent permeability, initial gas composition, and fracture length on gas production under various intrinsic permeability conditions.

MATHEMATICAL MODEL

PEBI Grid

PEBI grids are also known as Voronoi cells and are defined as the region in which all points are closer to the corresponding seed than any other seeds [18–20]. The boundary of each Voronoi cell is normal to the line connecting the seeds on the two sides. Compared to structured Cartesian and Corner gird, the unstructured PEBI grids have the following advantages.

- Flexibility: it can represent the characteristic of complex boundary, pinch-out, and faults in the formation precisely.

- Easiness of local refinement: because of the flexibility of PEBI grid and the arbitrariness of arranging PEBI grid point, it is easier to refine grid in the local place, such as domain around wells.

- Less grid orientation effect: the unstructured hexagonal PEBI grid makes the grid orientation effect less significant than structured grid.

- Easiness of discretizing and solving equations: the local orthogonality of PEBI grid makes it easier to discretize and solve equations by using finite volume method.

The PEBI grid used in this paper is based on previous researches [21, 22]. The grid points are arranged following streamline and based on well type, location, and reservoir geometry. The generated grids are denser near wells and looser far away from wells, as shown in Figure 1. This arrangement of PEBI grids can keep computation accuracy and save computation time.

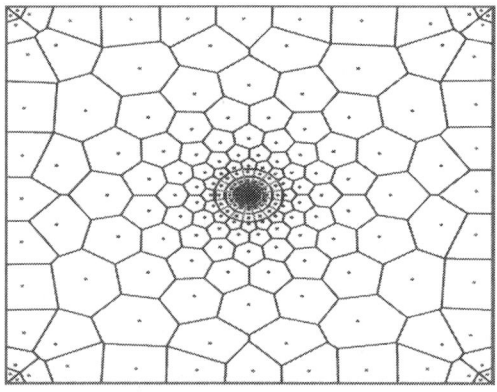

(a) One vertical well in the middle

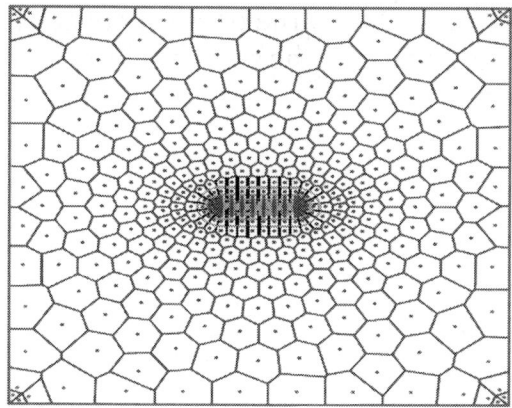

(b) One vertical well with a hydraulicfracture in the middle

Figure 1: Schematic of PEBI grid.

Apparent Permeability

Gas flow in low-permeability tight and shale gas reservoirs occurs following various mechanisms, such as slip flow, transition flow, and free molecular flow. The matrix permeability in such reservoirs needs to be modified to enable traditional Darcy's law describe such non-Darcy flow.

Note that the non-Darcy flow in the paper includes slip flow, transition flow, and free molecular flow which is defined as below.

Chambre and Schaaf [23] have classified four flow regimes based on Knudsen number (K_n), as shown in Table 1.

Table 1: Flow regimes classified by Chambre and Schaaf

K_n	>10.0	(0.1, 10)	$(10^{-3}, 0.1)$	$<10^{-3}$
Flow regimes	Free molecular flow	Transition flow	Slip flow	Continuum flow

The Knudsen number K_n expresses the mean free path of molecules as a fraction of a representative path (mean hydraulic radius, e.g.) [24]:

$$K_n = \frac{\lambda_g}{R_h}.$$

(1)

Here, l_g is the mean free path for gas and is defined by the following equation:

$$\lambda_g = \frac{\mu_g}{P} \sqrt{\frac{\pi RT}{2M_g}},$$

(2)

Where m_g is the gas viscosity in Pa·s, P is the absolute gas pressure in Pa, R is the universal gas constant (8,314 J/ (kmol·K)), T is the absolute temperature in Kelvin, and Mg is the molecular weight of the gas in kg/kmol.

The mean hydraulic radius of flow tubes in porous media R_h is defined as

$$R_h = 2\sqrt{2\tau} \sqrt{\frac{k_0}{\phi}},$$

(3)

where τ is the tortuosity, ϕ is the porosity, and k_0 is the intrinsic permeability of the reservoir in m².

Based on the unified model for gas flow in microtubes derived by Beskok and Karniadakis [4] and Florence et al. [5] proposed a formulation of apparent permeability in terms of Knudsen number to characterize the non-Darcy flow in the porous media. Consider

$$k = k_0 \left(1 + \alpha K_n\right) \left(1 + \frac{4K_n}{1 + K_n}\right)$$

(4)

where α is the dimensionless rarefaction coefficient and is defined by Beskok and Karniadak is as

$$\alpha = \frac{128}{15\pi^2} \tan^{-1}\left(4K_n^{0.4}\right) \tag{5}$$

Substituting (5) into (4) yields

$$k = k_0 \left(1 + \frac{128}{15\pi^2} \tan^{-1}\left(4K_n^{0.4}\right) K_n\right)\left(1 + \frac{4K_n}{1+K_n}\right) \tag{6}$$

This equation is valid for all flow regimes for gas flow in porous media [4, 5].

Multicomponent Langmuir Isotherm

To simulate and distinguish sorption capacity for different components, the extended Langmuir isotherm which is widely accepted by petroleum industry is used.

For component i the sorption volume is as follows:

$$V_{ads,i} = V_{L,i} \frac{y_i P}{P_{L,i}\left(1 + \sum_{j=1}^{n_h} y_j \left(P/P_{L,j}\right)\right)}, \quad i = 1,\ldots,n_h, \tag{7}$$

Where $V_{ads,}$ is the standard volume of sorbed component i, y_i is the mole fraction of the component i, and n_h is the total number of components. The Langmuir volume V_{Li} and Langmuir pressure P_{Li} are measured values for the pure component i. The total sorption is given by

$$V_{ads} = \sum_{i=1}^{n_h} V_{ads,i} = \sum_{i=1}^{n_h} V_{L,i} \frac{y_i P}{P_{L,i}\left(1 + \sum_{j=1}^{n_h} y_j \left(P/P_{L,j}\right)\right)} \tag{8}$$

Mass Conservation Equations

Based on finite volume method, the governing mass balance equation for the component i considering gas sorption is given by

$$\frac{\partial}{\partial t}\left(V\phi\rho_g y_i + V\rho_s V_{ads,i}\rho_{g,std}\right)$$

$$= \sum_l \left(T_r \frac{1}{\mu_g}\rho_g y_i \Delta\Phi_g\right)_l - \rho_{g,std} y_i q_{g,std},$$

$$i = 1, \ldots, n_h, \tag{9}$$

where V is the gas volume in m^3, ρs is the rock density in kg/m^3, ρ g,std is the gas density under standard conditions (1 atm and 15°C) in mol/m^3, ρg is the gas density under formation condition in mol/m^3, and $q g$,std is the gas production rate under standard condition in m^3/s. Values of ρg and ρg,std were calculated using the Peng-Robinson equation of state (PR EOS) [25]. The symbol l represents connection between adjacent grids, and $\Delta\Phi g$ is the difference in gas potentials between adjacent grids l_1 and l_2 and is given by $\Delta\Phi$ $g = \Phi_{l1} - \Phi_{l2} = p_{l1} - p_{l2} - (\rho_{l1} Z_{l1} - \rho_{l2} Z_{l2})$, where Z is the depth of the formation. The viscosity μg was calculated using the Lohrenz-Bray-Clark (LBC) correlation [26]. The transmissibility parameter T_r $= kA/L$, where k is the apparent permeability represented by (6), A is the cross-sectional area between the adjacent grids, and L is the distance between the adjacent grids.

The left term of (9) is the mass accumulation term and includes the mass of both free gas and sorbed gas. The first right termdenotes the advection term. The second right term denotes source or sink in the well, which stands for the rate of gas mass produced from or injected into a well.

For vertical well, the well production rate $q_{g, std}$ is expressed by Peaceman model:

$$q_{g,std} = \frac{\theta k h}{\mu_g \ln(r_e/r_w)} \left(p_j - p_{wf}\right) \tag{10}$$

where h is the effective height of well perforation in m, r_w is wellbore radius in m, p_j is well grid pressure in Pa, p_{wf} is bottom-hole flowing pressure in Pa, and r_e is the equivalent radius in the well grid in m. For structured Cartesian grid, r_e can be expressed as

$$r_e = 0.28 \frac{\left[\left(k_y/k_x \right)^{0.5} \Delta x^2 + \left(k_x/k_y \right)^{0.5} \Delta y^2 \right]^{0.5}}{\left(k_y/k_x \right)^{0.25} + \left(k_x/k_y \right)^{0.25}} \tag{11}$$

For unstructured PEBI grid, we derive r_e as follows.

Assume pressure at equivalent radius r_e as p_e which is equivalent to well grid pressure p_j. Arranging (10) yields

$$\left(p_e - p_{wf} \right) = \frac{\mu_g \ln \left(r_e/r_w \right) q_{g,\text{std}}}{\theta k h} \tag{12}$$

The flux from adjacent grids to well grid Q is the same as $q_{g,\text{std}}$ can be expressed as

$$Q = q_{g,\text{std}} = \sum_l \frac{T_r}{\mu_g} \left(p_l - p_j \right) = \sum_l \frac{T_r}{\mu_g} \left(p_l - p_e \right) \tag{13}$$

Arranging (10) and (12) and substituting into (13) yield

$$\ln r_e = \frac{\sum_l T_r \ln \Delta L - \theta k h}{\sum_l T_r}. \tag{14}$$

Consequently, r_e can be expressed as follows:

$$r_e = \exp \left(\frac{\sum_l T_r \ln \Delta L - \theta k h}{\sum_l T_r} \right) \tag{15}$$

For vertical well with a hydraulic fracture, we use the infinite conductivity model [27] to represent the conductivity of the fracture in the reservoir. Considering the wellbore storage, the well production rate $q_{g,\text{std}}$ can be expressed as

$$q_{g,\text{std}} = \frac{\sum_l \left(\left(T_r/\mu_g \right) \rho_g \Delta \Phi_g \right)_l}{\rho_{g,\text{std}}} - \frac{C}{\Delta t} \left(p_{wf}^{t_{n+1}} - p_{wf}^{t_n} \right) \tag{16}$$

where C is thewellbore storage in m³/MPa and t^{n+1} and t^n are $(n + 1)$th and nth time steps.

The unknown variables include pressure p, bottom-hole flowing pressure p_{wf}, and the mole fraction y_i in mass conservation equations (9). The nonlinear equations are solved using Newton iteration method and the matrix solver GMRES (generalized minimal residual method) [28].

MODEL VALIDATION

The model was validated by reproducing the pressure build-up data from a hydraulic fractured vertical well in a tight gas reservoir in Xinjiang, China. The pilot test operation comprised two stages: gas production at the rate of 8.36×10^4 m³/day for 30 days, followed by a shut-in period of 25 days.

The model was set up using parameters and properties listed in Table 2. The initial reservoir gas contains 94.7% CH_4, 1.2% CO_2, 2.7% C_2H_6, and 1.4% C_3H_8. The average gas sorption parameters (V_L and P_L) in Table 2 were provided without distinguishing different type of gas components. Figure 2 shows the predicted pressure dropdown during the production stage and the pressure build-up during the shut-in period. The modeling output was compared to the monitored high frequency/high-resolution BHP data during the shut-in operation [29]. The pressure data during the production was not available. The simulation output shows reasonable reproduction of the pressure build-up data.

Table 2: Properties and parameters of a tight gas reservoir inXinjiang, China

Name	Value	Units
Reservoir dimensions	3500 × 2000 × 18	Meter
Fracture half-length	230	Meter
Permeability	0.056	mD

Porosity	0.0376	N/A
Reservoir temperature	355.0	Kelvin
Initial pressure	31.5	MPa
Bulk density	2500.0	kg/m³
Wellbore diameter	0.084	meter
Langmuir volume (V_l)	0.0012	m³/kg
Langmuir pressure (P_l)	7.3	MPa

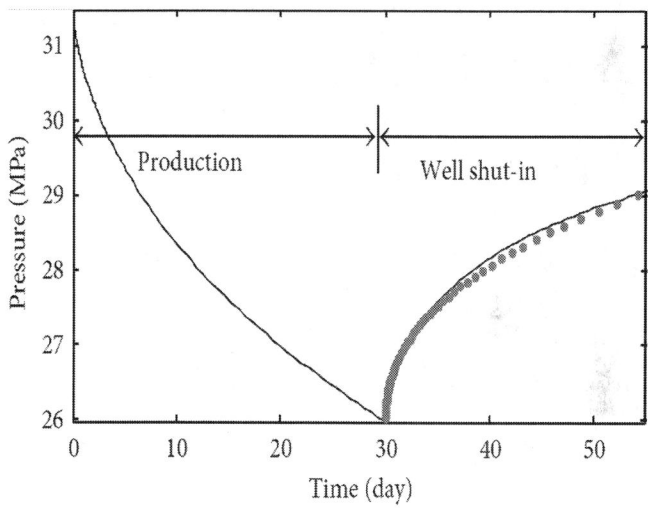

Figure 2: The bottom-hole pressure history of a production well in a tight gas reservoir in Xinjiang, China. The production rate from day 0 to 30 was 8.36×104 m3/day, where pressure data was not available. The predicted pressure build-up (black line) during the shut-in period compares well with data (red symbol).

RESULTS AND DISCUSSION

The developed model was used to understand the effect of non-Darcy flow, length of the hydraulic fracture, and initial gas composition on gas production.

Specific reservoir parameters and simulation conditions are listed in Table 3, unless noted otherwise. The sorption parameters for gas species are listed in Table 4. The hydraulically fractured vertical well is located in the middle of simulated gas reservoir of which the boundary is sealed.

Table 3: Input parameters for simulated hydraulically fractured vertical well and the gas reservoir

Input parameter	Values	Unit
Dimension	2000 × 1000 × 25	m
Fracture half-length	200	m
Wellbore storage	1	m³/MPa
Flowing bottom-hole pressure	10	MPa
Porosity	0.05	
Temperature	60	°C
Initial reservoir pressure	35	MPa
Initial gas composition	2% CO_2, 90% CH_4, 8% C_2H_6	
Bulk density	2500	Kg/m³

Table 4: Langmuir constants for gas species in gas reservoir

	CO_2	CH_4	C_2H_6
V_l (m³/t)	4.1	1.6	2.6
P_l (MPa)	5.76	10.77	5.59

The Effect of Non-Darcy Flow on Gas Production

Here, the gas production was calculated with and without considering non-Darcy flow under three different intrinsic permeability conditions which are 1.01 × 10⁻² mD (10^{-17} m²), 1.01

$\times 10^{-4}$ mD (10^{-19} m²), and 1.01×10^{-6} mD (10^{-21} m²), respectively.

The non-Darcy flow is incorporated into apparent permeability which is represented by (6). The non-Darcy flow in tight formations includes slip flow, transition flow, and free molecular flow which will enhance the gas flow ability, especially under lower pressure condition. Thus, as gas produces, the reservoir pressure continues to decrease and the influence of non-Darcy flow on gas production will become more significant.

In the higher permeable reservoir ($k_0 = 1.01 \times 10^{-2}$ mD), the influence of non-Darcy flow on gas production is very limited, as depicted by Figure 3(a). The flow regime of gas mainly stays in slip flow regime. After 4.38×104 days (120 years) production, the difference of predicted produced gas volume with and without considering non-Darcy flow is only about 2%, as shown in Figure 4.

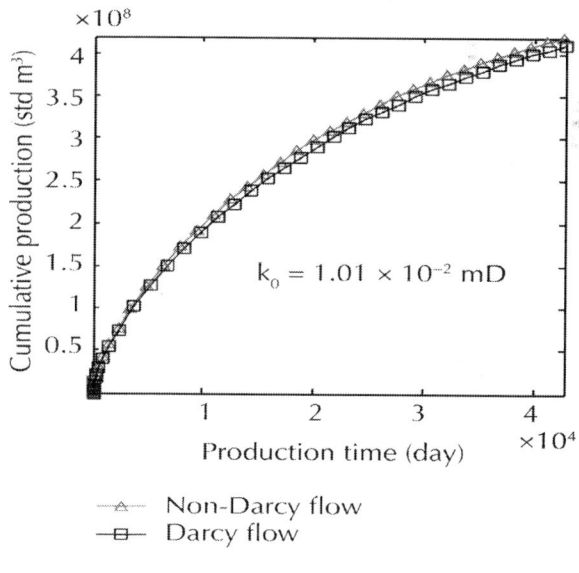

(a)

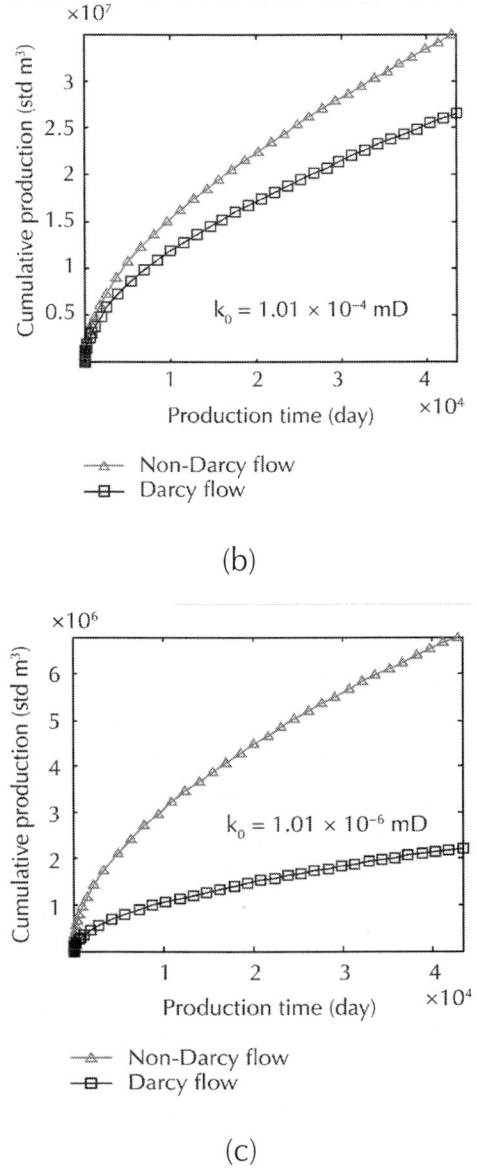

(b)

(c)

Figure 3: Thegas production predicted with and without considering non-Darcy flow under three different intrinsic permeability conditions.

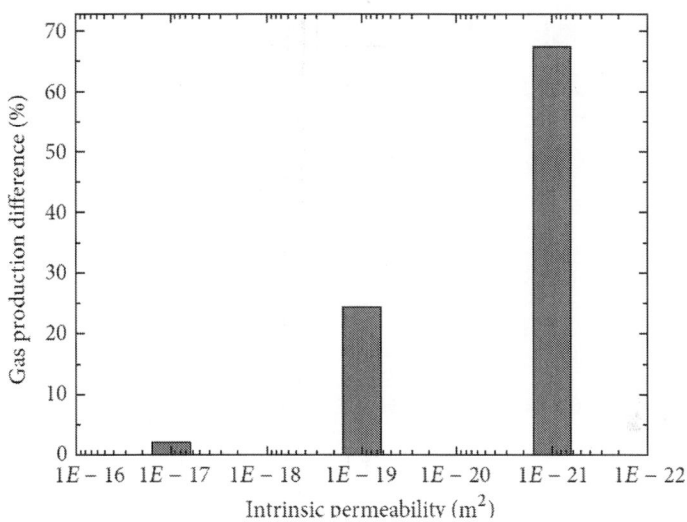

Figure 4: Gas production difference predicted with considering non-Darcy flow and without considering it after 120 years of production.

While in lower permeable gas reservoirs (k_0= 1.01 ×10^{-4} mD, k_0= 1.01 × 10^{-6} mD) which are more common for tight/shale gas reservoirs, the non-Darcy flow (mainly under transition flow regime) reveals more significant influence on produced gas volume, as depicted in Figures 3(b) and 3(c). The predicted produced gas volume without considering non-Darcy flow is 24% and 67% smaller than that with considering non-Darcy flow for reservoir k_0= 1.01×10^{-4} mD and reservoir k_0= 1.01 × 10^{-6} mD, respectively.

Lower permeability often indicates smaller pore diameter which means that more chance of non-Darcy flow would happen during gas flow. From Figures 3(a) to 3(c), as the permeability goes smaller, the gas flow regimes mainly stay on slip flow, early transition flow, and late transition flow, respectively.

Based on the results above, non-Darcy flow is critical for accurate predicting gas flow behavior in low permeable reservoirs, such as tight gas or shale gas reservoirs. Ignoring it would significantly underestimate gas production capacity.

The Effect of Length of Hydraulic Fracture on Gas Production

Here, we aim to understand the role of hydraulic fracture length in determining gas production. Gas production rate and cumulative gas production were predicted when the half-length of hydraulic fracture equals 0, 50, 100, 200, and 400 meters in the reservoir of which intrinsic permeability equals 1.01×10^{-2} mD, 1.01×10^{-4} mD, and 1.01×10^{-6} mD.

Take the medium permeable reservoir ($k_0 = 1.01 \times 10^{-4}$ mD), for example. The gas cumulative production and production rate are depicted in Figure 5.

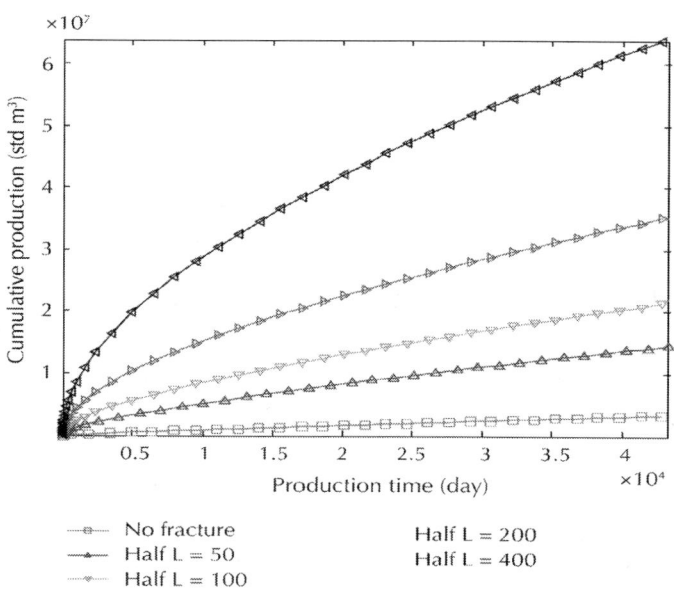

(a) Cumulative production

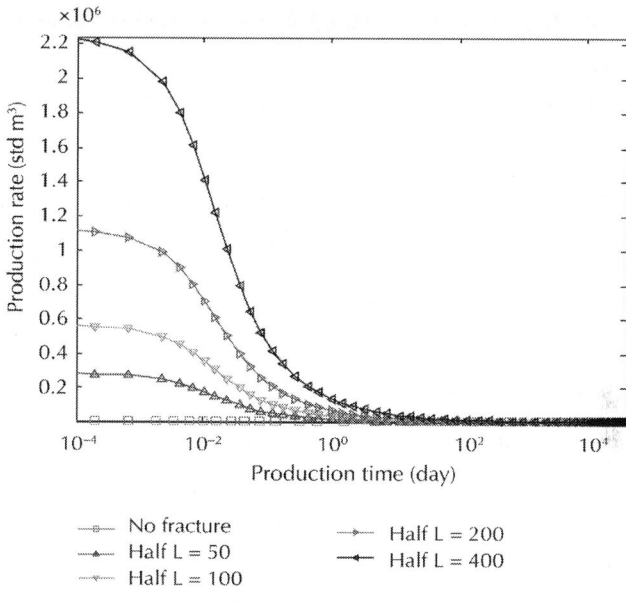

(b) Production rate

Figure 5: Effect of fracture half-length on predicted gas cumulative production and production rate.

The results show that gas production rate and cumulative production increase with half-length of hydraulic fracture. The vertical well with no fracture (represented by red line) indicates very limited production potential. The averaged production rate is only 79 m³/day and the cumulative production after 120 years is only 3.4 million cubic meters, as shown in Figures 5(a) and 5(b), while for the vertical well with hydraulic fracture of which the half-length is only 50 meters, the cumulative production increases to 15 million cubic meters, about five times as that of the well with no fracture. As the half-length of fracture increases to 100 m, 200 m, and 400 m, the cumulative production increases to 22, 36, and 65 million cubic meters, respectively. With the increase of half-length of fracture, the extent of cumulative production increased is remarkable, indicating the significant importance of length of hydraulic fracture on gas production.

Gas production rate in such low permeable reservoir decreases very quickly, especially in the early production time, as depicted by Figure 5(b). For the well with a fracture half-length of 400 meters, in the first 2 days of production, the rate stays above 100,000 m³/day, while after 200 days of production, the rate decreases rapidly to 10,000 m³/day. In the late production time, the rate decreases much slower. After 100 years of production, the rate remains fairly at 1,000 m³/day.

For all simulated reservoirs, the increase of half-length of hydraulic fracture shows a positive influence on gas production, as depicted in Figure 6. In the higher permeable reservoir (k_0= 1.01 × 10^{-2} mD), the cumulative production shows nonlinear relationship with fracture half-length. With increase of fracture half-length, the increment of cumulative production decreases gradually and may finally become negligible. In the medium permeable reservoir (k_0= 1.01 ×10^{-4} mD), the cumulative production shows almost linear increment relationship with fracture half-length. And in the lower permeable reservoir (k_0= 1.01 ×10^{-6} mD), the relationship between the two can be identified as linear.

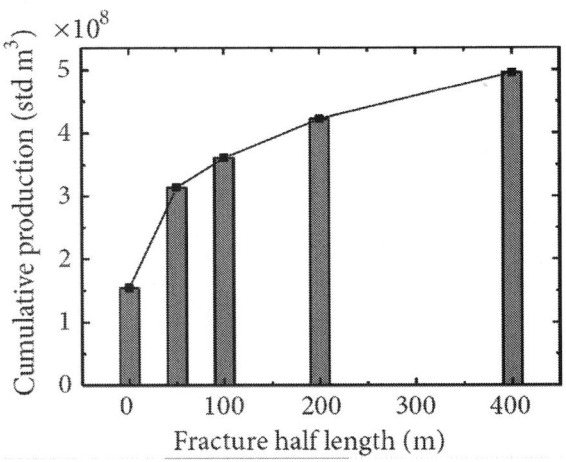

(a) k_0= 1.01 × 10^{-2} mD

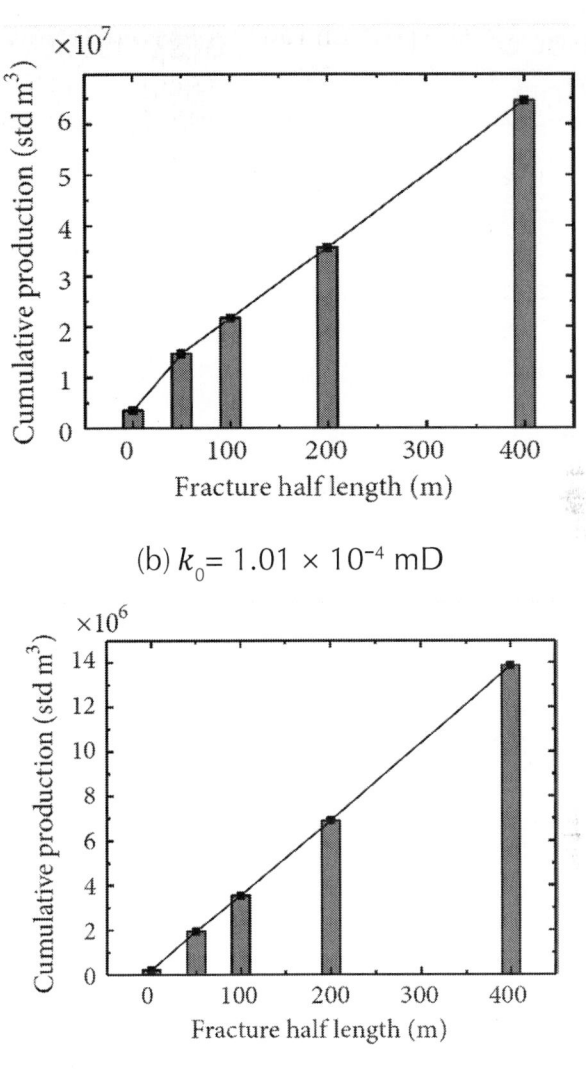

(b) $k_0 = 1.01 \times 10^{-4}$ mD

(c) $k_0 = 1.01 \times 10^{-6}$ mD

Figure 6: Effect of fracture half-length on cumulative production after 120 years of production in three different permeable gas reservoirs.

Thus, we conclude that the lower the permeability is, the more significant the increasing fracture length influences gas production. But for some relatively higher permeable reservoirs, an appropriate length of fracture should be found considering the balance between economical cost of fracturing a well and gas production.

The Effect of Initial Gas Composition on Gas Production

Here, gas cumulative productions are calculated when initial CO_2 composition equals to 0%, 10%, 20%, and 30%, respectively. The half-length of hydraulic fracture is confined to 200 meters. The intrinsic permeability is confined to 1.01×10^{-4} mD. The initial gas species in the reservoir is assumed to be CO_2 and CH_4. All other parameters are the same as in Tables 3 and 4. The results are shown in Figure 7.

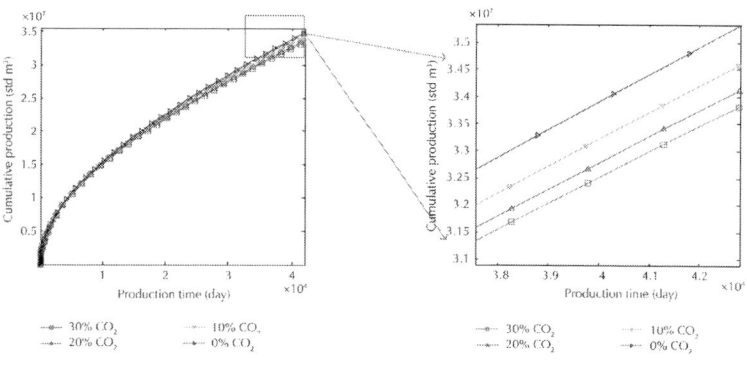

Figure 7: Effect of initial gas composition on cumulative production.

The results show initial gas composition has some influence on cumulative production. Higher percentage of CO_2 in the initial gas leads to lower gas production which can be seen clearly on the enlarged figure. This is mainly because the presence of CO_2 increases gas viscosity, which would slow down flow speed in the formation and eventually reduce gas production.

The decrement of cumulative gas production slows down with the increase of CO_2 percentage in initial gas content, as depicted in Figure 8(a), while that of cumulative methane production does not, as depicted in Figure 8(b). The cumulative methane production shows a linear relationship with the percentage of CO_2 in initial gas content. When initial CO_2 percentage equals 0%, the produced cumulative gas volume is 4.4% (1.5 million m³) higher than that of the case when initial CO_2 percentage equals 30%, while for produced cumulative methane volume, the former is 49.8% (11.9 million m³) higher than the latter.

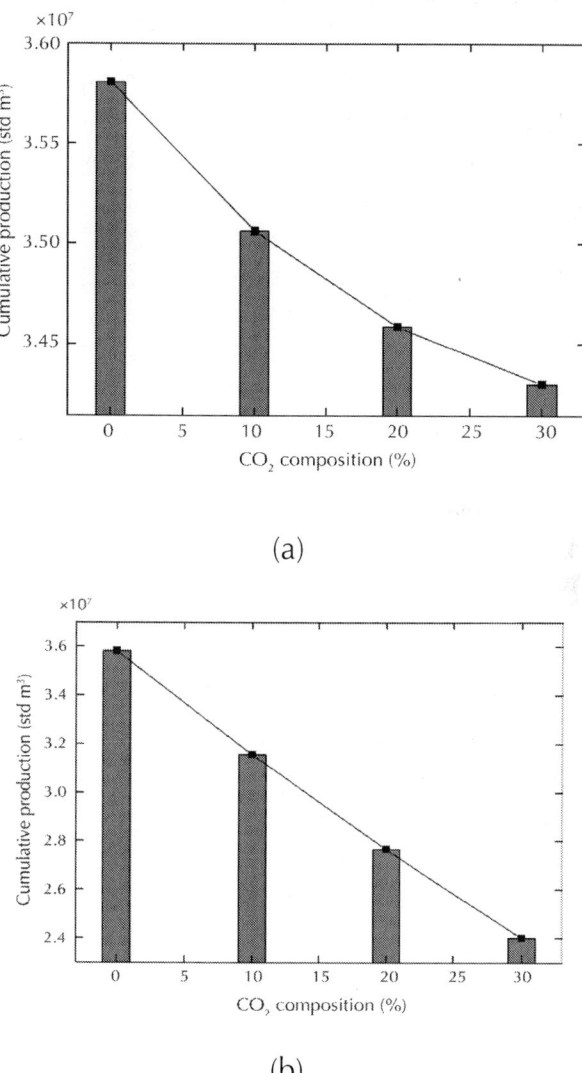

(a)

(b)

Figure 8: Effect of initial gas composition on produced (a) cumulative mixed gas volume and (b) methane volume after 120 years of production.

CONCLUSIONS

In this paper, we proposed a compositional model for tight/shale gas reservoirs based on unstructured PEBI grid. The non-Darcy flow, including slip flow, transition flow, and free molecular flow, is considered in terms of apparent permeability in the model. Multicomponent adsorption is also considered in terms of extended Langmuir isotherm.

With the proposed model, we studied the effect of non-Darcy flow, length of the hydraulic fracture, and initial gas composition on gas production. The results showed the following: (1) the non-Darcy flow shows significant influence on gas production, especially in low permeable reservoirs. Ignoring this effect would lead to 67% underestimation on produced cumulative gas volume according to the simulated case. (2) Gas production increases with half-length of hydraulic fracture. However, in higher permeable reservoirs, the increment of gas production decreases with increase in fracture length, while in lower permeable reservoirs, gas production almost increases linearly with fracture length. Gas production would impossibly reach to economic rate without fracturing the well. (3) Higher initial CO_2 percentage would lead to lower gas production, for its ability to increase gas viscosity in the formation.

With considering non-Darcy flow and multicomponent gas adsorption in tight/shale gas reservoirs, the proposed compositional model can be a powerful tool to predict gas behavior in such unconventional reservoirs and be used to offer valuable insights into reservoir engineers to make better exploitation schemes.

ACKNOWLEDGMENTS

The authors are grateful for funding from Major State Basic Research Development Program of China (973 Program) (no. 2011CB707305) and National Key Science and Technology Project (2011ZX05009-006).

REFERENCES

1.   P. H. Nelson, "Pore-throat sizes in sandstones, tight sandstones, and shales," AAPG Bulletin, vol. 93, no. 3, pp. 329–340, 2009. View at Publisher

2.   F. Javadpour, "Nanopores and apparent permeability of gas flow in mudrocks (shales and siltstone),"Journal of Canadian Petroleum Technology, vol. 48, no. 8, pp. 16–21, 2009.

3.   F. Javadpour, D. Fisher, and M. Unsworth, "Nanoscale gas flow in shale gas sediments," Journal of Canadian Petroleum Technology, vol. 46, no. 10, pp. 55–61, 2007.

4.   A. Beskok and G. E. Karniadakis, "A model for flows in channels, pipes, and ducts at micro and nano scales," Microscale Thermophysical Engineering, vol. 3, no. 1, pp. 43–77, 1999. View at Publisher ·

5.   F. Florence, J. Rushing, K. E. Newsham, and T. A. Blasingame, "Improved permeability prediction relations for low permeability sands," in Proceedings of the Rocky Mountain Oil & Gas Technology Symposium, 2007.

6.   F. Civan, "Effective correlation of apparent gas permeability in tight porous media," Transport in Porous Media, vol. 82, no. 2, pp. 375–384, 2010.

7.   F. Civan, C. S. Rai, and C. H. Sondergeld, "Shale-gas permeability and diffusivity inferred by improved formulation of relevant retention and transport mechanisms," Transport in Porous Media, vol. 86, no. 3, pp. 925–944, 2011.

8.   Q. Zheng and B. Yu, "A fractal permeability model for gas flow through dual-porosity media," Journal of Applied Physics, vol. 111, no. 2, Article ID 024316, 2012.

9.   Q. Zheng, B. Yu, S. Wang, and L. Luo, "A diffusivity model for gas diffusion through fractal porous media," Chemical Engineering Science, vol. 68, no. 1, pp. 650–655, 2012.

10. Q. Zheng, B. Yu, Y. Duan, and Q. Fang, "A fractal model for gas slippage factor in porous media in the slip flow regime,"

Chemical Engineering Science, vol. 87, pp. 209–215, 2013. View at Publisher ·View at Google Scholar

11. C. M. Freeman, G. J. Moridis, and T. A. Blasingame, "A numerical study of microscale flow behavior in tight gas and shale gas reservoir systems," Transport in Porous Media, vol. 90, no. 1, pp. 253–268, 2011.

12. C. M. Freeman, G. J. Moridis, and T. A. Blasingame, "Modeling and performance interpretation of flowing gas composition changes in shale gas wells with complex fractures," in Proceedings of the International Petroleum Technology Conference: Challenging Technology and Economic Limits to Meet the Global Energy Demand (IPTC '13), pp. 4868–4883, Beijing, China, March 2013.

13. C. R. Clarkson, M. Nobakht, D. Kaviani, and T. Ertekin, "Production analysis of tight-gas and shale-gas reservoirs using the dynamic-slippage concept," SPE Journal, vol. 17, no. 1, pp. 230–242, 2012.

14. T. Ertekin, G. R. King, and F. C. Schwerer, "Dynamic gas slippage: a unique dual-mechanism approach to the flow of gas in tight formations," SPE Formation Evaluation, vol. 1, no. 1, pp. 43–52, 1986.

15. V. Swami, F. Javadpour, and A. Settari, "A numerical model for multi-mechanism flow in shale gas reservoirs with application to laboratory scale testing," in Proceedings of the 75th EAGE Conference & Exhibition Incorporating SPE EUROPEC, 2013.

16. J. Li, C. Wang, D. Ding, Y. S. Wu, and Y. Di, "A generalized framework model for simulation of gas production in unconventional gas reservoirs," in Proceedings of the SPE Reservoir Simulation Symposium, The Woodlands, Tex, USA, February 2013. View at Publisher ·

17. J. Yao, H. Sun, D.-Y. Fan, C.-C. Wang, and Z.-X. Sun, "Numerical simulation of gas transport mechanisms in tight shale gas reservoirs," Petroleum Science, vol. 10, no. 4, pp. 528–537, 2013. View at Publisher ·

18. Z. E. Helnemann, C. W. Brand, M. Munka, and Y. M. Chen,

"Modeling reservoir geometry with irregular grids," SPE Reservoir Engineering, vol. 6, no. 2, pp. 225–232, 1991.

19. Z. E. Heinemann and C. W. Brand, "Gridding techniques in reservoir simulation," in Proceedings of the 1st International Forum on Reservoir Simulation, Alpbach, Austria, 1988.

20. C. L. Palagi and K. Aziz, "Use of Voronoi grid in reservoir simulation," SPE Advanced Technology Series, vol. 2, no. 2, pp. 69–77, 1994. View at Publisher

21. L. Zhang, D. Li, W. Zha, et al., "Generation and application of adaptive PEBI grid for numerical well testing(NWT)," in Proceedings of the International Conference on Mechatronic Sciences, Electric Engineering and Computer (MEC '13), pp. 3002–3006, Shenyang, China, 2013.

22. W. Zha, Numerical Reservoir Calculation on PEBI Grid and Implementation, University of Science and Technology of China, 2009.

23. P. A. Chambre and S. A. Schaaf, Flow of Rarefied Gases, 1961.

24. L. B. Loeb, The Kinetic Theory of Gases, Courier Dover, 2004.

25. S. M. Walas, Phase Equilibria in Chemical Engineering, vol. 4, Butterworth, Boston, Mass, USA, 1985.

26. J. Lohrenz, B. Bray, and C. Clark, "Calculating viscosities of reservoir fluids from their compositions,"Journal of Petroleum Technology, vol. 16, no. 10, pp. 1171–1176, 1964.

27. B. Horsfield and H. M. Schulz, "Shale gas exploration and exploitation," Marine and Petroleum Geology, vol. 31, no. 1, pp. 1–2, 2012. View at Publisher

28. Y. Saad and M. H. Schultz, "GMRES: a generalized minimal residual algorithm for solving nonsymmetric linear systems," SIAM Journal on Scientific and Statistical Computing, vol. 7, no. 3, pp. 856–869, 1986.

29. W. Ding, C. Li, C. Li et al., "Fracture development in shale and its relationship to gas accumulation,"Geoscience Frontiers, vol. 3, no. 1, pp. 97–105, 2012.

Chapter 4

Formation Pressure Testing at the Mount Elbert Gas Hydrate Stratigraphic Test Well, Alaska North Slope: Operational Summary, History Matching, and Interpretations

Brian Anderson[a,b], Steve Hancock[c], Scott Wilson[d], Christopher Enger[e], Timothy Collett[f], Ray Boswell[a], and Robert Hunter[g]

[a]National Energy Technology Laboratory, 3610 Collins Ferry Road, Morgantown, WV 26507, USA

[b]West Virginia University, Department of Chemical Engineering, Morgantown, WV 26506-6102, USA

cRPS Energy Canada, 1400, 800 Fifth Ave. SW, Calgary, Alberta, Canada T2P 3T6

dRyder Scott Company, Petroleum Consultants, 621, 17th Street, Suite 1550, Denver, CO 80293, USA

eRock Mechanics Laboratory, Colorado School of Mines, 1500 Illinois St, Golden, CO 80401, USA

fUS Geological Survey, Denver Federal Center, MS-939, Box 25046, Denver, CO 80225, USA

gASRC Energy Services, 3900 C Street, Suite 702, Anchorage, AK 99503, USA

ABSTRACT

In February 2007, the U.S. Department of Energy, BP Exploration (Alaska), and the U.S. Geological Survey, collected open-hole pressure-response data, as well as gas and water sample collection, in a gas hydrate reservoir (the BPXA-DOE-USGS Mount Elbert Gas Hydrate Stratigraphic Test Well) using Schlumberger's Modular Dynamics Formation Tester (MDT) wireline tool. Four such MDT tests, ranging from six to twelve hours duration, and including a series of flow, sampling, and shut-in periods of various durations, were conducted. Locations for the testing were selected based on NMR and other log data to assure sufficient isolation from reservoir boundaries and zones of excess free water. Test stages in which pressure was reduced sufficiently to mobilize free water in the formation (yet not cause gas hydrate dissociation) produced readily interpretable pressure build-up profiles. Build-ups following larger drawdowns consistently showed gas-hydrate dissociation and gas release (as confirmed by optical fluid analyzer data), as well as progressive dampening of reservoir pressure build-up during sequential tests at a given MDT test station.

History matches of one multi-stage, 12-h test (the C2 test) were accomplished using five different reservoir simulators: CMG-STARS, HydrateResSim, MH21-HYDRES, STOMP-HYD, and TOUGH + HYDRATE. Simulations utilized detailed information collected across the reservoir either obtained or determined from geophysical well logs, including thickness (11.3 m, 37 ft.), porosity (35%), hydrate saturation (65%), both mobile and immobile water saturations, intrinsic permeability (1000 mD), pore water salinity (5 ppt), and formation temperature (3.3–3.9 °C). This paper will present the approach and preliminary results of the history-matching efforts, including estimates of initial formation permeability and analyses of the various unique features exhibited by the MDT results.

INTRODUCTION

The DOE-BPXA-USGS Mount Elbert Gas Hydrate Stratigraphic Test Well (Mount Elbert Well) was drilled, cored and tested in February 2007 as part of a government-industry cooperative program to evaluate the resource potential of gas hydrate on the North Slope of Alaska (Boswell et al., 2008 and Hunter et al., 2011), see Fig. 1. Key among the goals of the project were to collect various data in the field to support the development of numerical models of gas hydrate reservoirs in advance of planning for extended term production testing. In addition to a full suite of core and wireline log data, a program of open-hole pressure-response tests was conducted using Schlumberger's Modular Dynamic Tester (MDT) tool. These tests were designed to provide insight into near-wellbore response to pressure declines that could be used to deduce key reservoir petrophysical parameters.

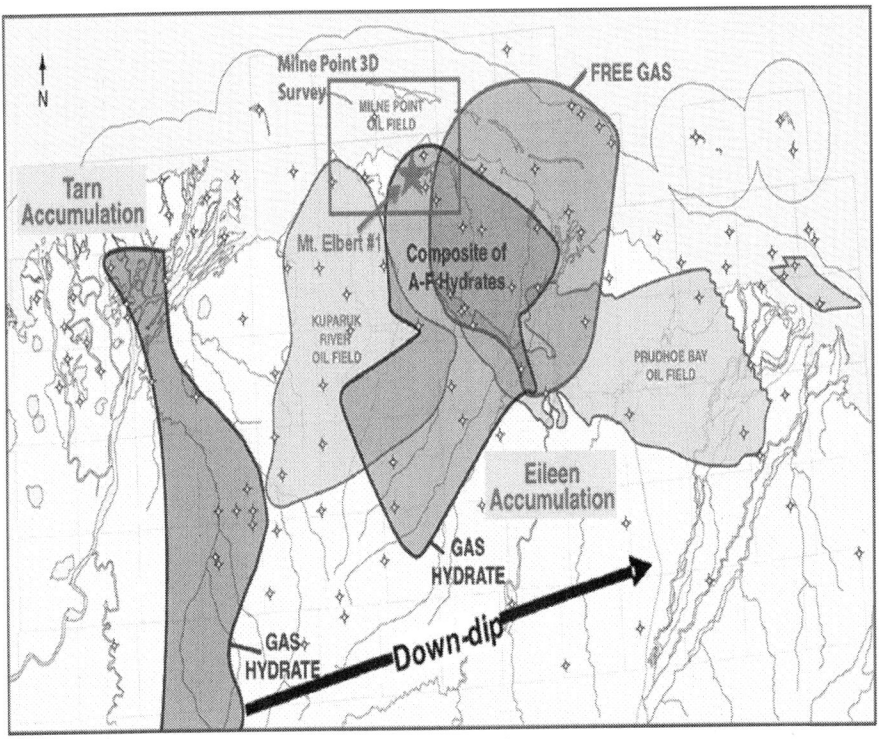

Figure 1: Location of delineated gas-hydrate prospects and the Alaskan North Slope (after (Inks et al., 2009)).

Four MDT tests, each containing a series of flow and shut-in periods of varying length, were conducted at four stations zones in two different gas hydrate-bearing sand reservoirs. Each of these four tests, denoted C1, C2, D1, and D2, were conducted in hydrate-bearing sediments as shown in Fig. 2. The characteristics of these four tests are shown in Table 1.

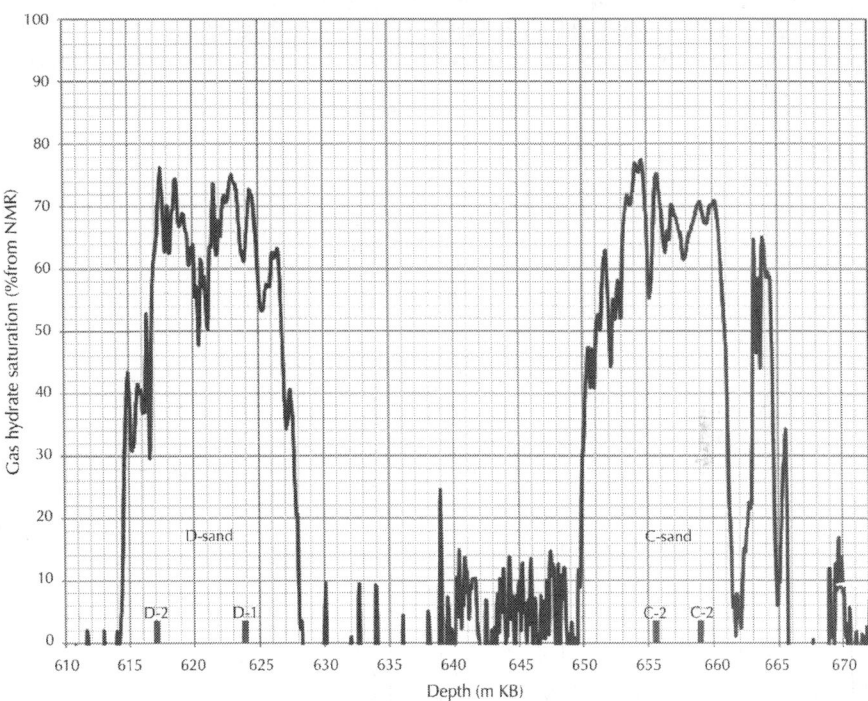

Figure 2: Gas-hydrate saturation with depth based on magnetic resonance log data from the Mount Elbert stratigraphic test well showing the location of the four MDT tests within the C and D sand units.

Table 1: MDT test intervals

Test no.	Type	MDT intake depth (m)	Hydrate saturation (%)	Pressure (MPa)			Temp (°C)
				Mud	Pore	Stability	
C1	Packer	658.7	70%	7.67	6.47	3.81	3.8
C2	Packer	655.6	75%	7.63	6.44	3.73	3.6
D1	Packer	623.9	70%	7.27	6.13	3.35	2.5
D2	Packer	617.2	78%	7.19	6.06	3.27	2.3

History of Gas Hydrate MDT Testing

The first direct production test of naturally occurring gas hydrates using pressure depletion occurred in 1972, at both the Imperial Oil Limited Mallik L-38 and Ivik J-26 wells, located on the MacKenzie Delta in Canada's North West Territories (Bily and Dick, 1974). Both wells were tested using the closed chamber technique, where the formation flows into a fixed volume initially at low or atmospheric pressure. While no useful pressure transient analysis could be conducted with the data that was obtained, there was an apparent pressure response indicating that fluid (assumed gas and water) was produced into the wellbore.

The 2002 Japex et al Mallik 5L-38 production testing program (Dallimore and Collett, 2005) was designed to evaluate gas hydrate production through both pressure depletion and thermal stimulation. For the pressure depletion tests, the intent was to conduct tests similar to the 1972 closed chamber tests over multiple intervals. While closed chamber testing is still used occasionally, a more modern and efficient method to conduct small scale reservoir productivity tests and to obtain reservoir petrophysical, temperature, pressure and fluid data is through the use of Schlumberger's MDT® wireline-conveyed logging tool (Schlumberger, Date). The tool is generally deployed in open-hole conditions, even in unconsolidated sandstone reservoirs.

For the Mallik 5L-38 well, based on concerns for borehole stability and preservation of the gas hydrate, the well was cased prior to any testing operations. Six separate intervals, including a free gas zone at the base of the gas hydrate stability field, a water zone, and four different hydrate intervals were tested. Each zone was perforated using deep penetrating charges at 5 shots per foot, in order to maximize surface area contact with the gas hydrate. There were also concerns with respect to reservoir temperature. The MDT tool had not been operated to test in such a low temperature environment (Mallik formation temperatures ranged from 10 to 14 °C for the planned intervals). Inflatable packers were used at Mallik to straddle each planned test interval and isolate it from the wellbore. Based on the diameter of the production casing, the packers could be spaced a maximum of ~0.75 m apart, which limited the perforations for each planned test interval to 0.5 m. A dual-stage, double-acting, reversible, positive-displacement pump was used to set the packers, pump fluid out of and into the reservoir, and pressurize the sample containers. This module created the pressure drawdown to initiate gas hydrate dissociation. An optical flow-analyzer module was used to determine the relative abundance of oil, gas, and water-based fluids being pumped through the MDT tool. A pressure–temperature gauge records all data for pressure transient analysis purposes. The gauge is located below the packer module and therefore could not be used to monitor temperature changes in the flow stream associated with the endothermic dissociation process. The MDT tool used a Mallik featured a multichamber sample module that allowed collection of separate gas and water samples from the test intervals. The MDT tool was also equipped with gamma-ray and casing-collar-locator tools for depth-control purposes. A large, single-chamber sample tool was also pre-charged with methanol and nitrogen, which could be released if there was evidence of plugging due to formation of gas hydrate in the tool passages. The tool measured approximately 28 m length, and weighed over 1000 kg.

The original premise for the Mallik 5L-38 MDT tests in the gas hydrate-bearing intervals was to reduce reservoir pressure below the gas hydrate stability point, and to shut-in and observe the pressure build-up. It was anticipated that the rate of gas production would be too small to measure, and that the rate of gas hydrate dissociation would have to be inferred from changing pressure data. Two important phenomena were observed during the gas hydrate tests: after an initial clean-up flow of borehole fluids, natural gas with small or trace amounts of water was produced; and, upon shut-in, the pressure response displayed typical porous-media effects and indicated both flow contribution and pressure effects beyond the near-wellbore area. The MDT test procedures for the gas hydrate intervals were then modified to include multiple flow and build-up periods, as well as injection (mini-frac) and pressure fall-off periods.

Design of Mount Elbert MDT Testing Program

For the Mount Elbert testing, the MDT tool was configured similar to that used at Mallik as described above. Based on the experience at Mallik, dual pump out modules were incorporated into the Mount Elbert tool string to improve reliability in the event of fines production, which can cause pump wear and failure. This risk was particularly acute for the prolonged (12-h) tests planned for Mount Elbert: the Mallik tests were typically less than 3 h per test station, and operations there were disrupted by the need to pull the MDT tool from the hole to repair a plugged pump. Another addition to the Mount Elbert program was the incorporation of a several small pressure–temperature recorders within a hollow steel pipe welded to the outside of the pump intake screen. One of these devices survived the test, and provided temperature data with time that could be related to the various stages of the test. This tool did not have surface read-out as part of the MDT tool string, therefore the temperatures associated with the test results discussed in the following sections were not observed until after the MDT was recovered to surface.

The Mount Elbert MDT tool had two alternative fluid intake ports. In addition to the port located inside the intake screen, there was also a conventional small diameter probe and sealing pad that could be extended from the tool and set against the borehole to extract fluids. This option was attempted at Mount Elbert, but failed, as the sands in the Mount Elbert well were too soft to properly set the pad and obtain a seal. A schematic of the MDT tool is shown in Fig. 3

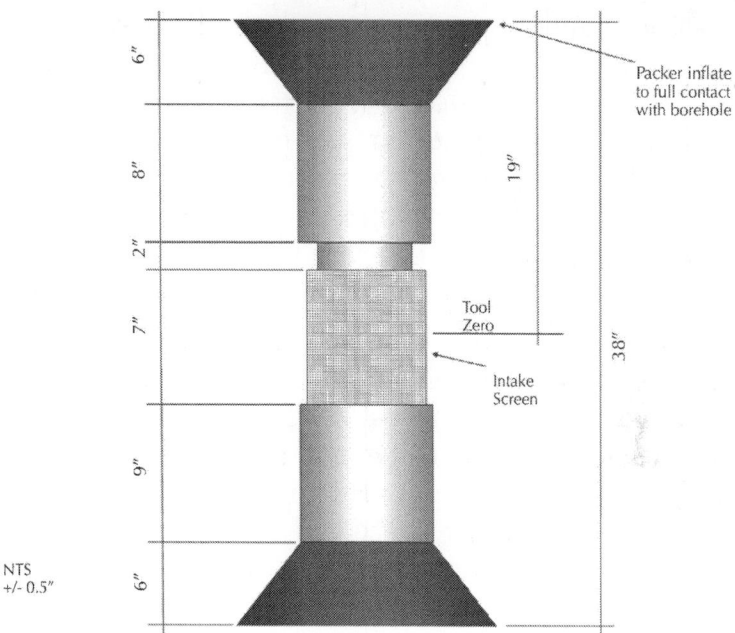

Figure 3: Schematic of the Schlumberger Modular Dynamic Tester used in the V Elbert Stratigraphic Test Well.

Operationally, there were two significant differences between the planned Mount Elbert tests and those conducted at Mallik. First, the Mount Elbert tests were conducted in an open hole, as opposed as through perforated casing. This eliminated the potential impact of perforating tunnels perhaps influencing the pressure transient data, resulting in much more complex data interpretation. Also, while there was no indication of sand production during the 2002 MDT test at Mallik, the longer planned testing times at Mount Elbert resulted in the decision to deploy the MDT tool on drillpipe to fully mitigate the risks of tool sticking due to fill. To further assess the potential for sticking the tool, the first test was limited to 6 h in duration in order that the tool could be moved to check for sticking. No significant overpull was noted at any of the test intervals after the packers were released. These actions slightly increased the time to run the tool to depth and move the tool between test intervals, but otherwise did not impact test results.

A second significant contrast with the earlier gas hydrate MDT testing at Mallik was the reservoir temperature. While Mallik reservoirs ranged from 7.5 to 13 °C, the temperatures of the planned test intervals at Mount Elbert ranged from approximately 2 to 3 °C. While these lower temperatures created additional concern over the potential to form hydrates inside the MDT tool, they also created additional concern over the formation of ice due to the endothermic nature of the hydrate dissociation process.

Test intervals at Mount Elbert were carefully selected using the NMR and other log data (Table 1). A primary criterion in selection of the sites was to isolate the test zones within zones of relatively homogenous reservoir character and gas hydrate saturation. In particular, test stations were located at least 1 m away from any potential excess free water that could negatively impact the test. Stations were also, as feasible, located away from vertical lithologic changes that could act of flow boundaries. In addition, the stations were vertically separated from one another to assure that reservoir within one test had not been impacted by a previous test. In addition, the physical separation of the test locations assured that the packers would not be set in a previously tested area where sediment disturbance could lead to difficulty in packer seal. For the C unit, which had an available test interval from 654.1 to 660.2 m (water below and poor reservoir quality above), test intervals were located at 657.1 to 660.2 m (C1 test) and 654.1 to 657.1 m (C2 test)

For the D unit, which had an available test interval from 616.3 to 625.8 m (water above and below), test stations were located at 622.4–625.5 m (D1 test) and 615.7–618.0 m (D2 test). This last station was located only ~0.3 m meters below a conglomeratic and reduced porosity zone at 617.5 m. The MDT tool configured for Mount Elbert had a packer section 2.7 m in overall length (2 × 0.9 m packer elements with 0.9 m spacing in between packers). Detailed descriptions of the lithology (Rose et al., 2011), the geochemistry (Torres et al., 2011), and the sediment physical properties (Winters et al., 2011) can be found in other papers within this issue.

THE MOUNT ELBERT MDT PROCEDURE

The objectives of the Mount Elbert MDT tests were to determine reservoir response to fluid withdrawal and pressure reduction; provide an indication of reservoir quality and performance; and obtain samples of produced water and gas. As such, the tests were planned to be conducted as typical tight gas MDT's with multiple flow and build-up periods, similar to the procedure used at Mallik. Injection tests were not planned.

The four tests included a series of stages that included continuous pressure and temperature monitoring during alternating periods of flow (pressure drawdown), and shut-in (pressure build-up) of various duration. Samples of produced fluids were also taken and the nature of produced fluids was continuously monitored using the tools optical fluid analyzer.

During the flow (pumping) periods, fluids (potentially containing a mixture of formation water and free methane gas) were extracted by the tool, thereby reducing the pressure in the formation in the vicinity of the well. Short-term MDT testing does not provide reliable information on reservoir deliverability or potential production rate, particularly in tight formations. However, by examining the recovery of the pressure within the formation after cessation of the withdrawal of fluids resulting from each flow period, it was hoped that key reservoir parameters associated with the formation could be extracted. The pressure and temperature were measured directly during the various flow and build-up periods of the MDT tests. Produced fluid volumes (aqueous, gas, and oil-based drilling fluid) were not measured directly, but were later estimated by Schlumberger from the stroke data for the positive-displacement pump in the MDT and the optical analyzer data, which provides an approximate measure of fluid volume ratios for each component. Without more detailed produced fluid volume data, the numerical simulation history matching was less constrained by the produced fluid volumes than the pressure and temperature measurements.

The C1 MDT Flow Test

The first MDT test, C-1, was planned to be a short-duration test. As seen in Fig. 4, at an MDT intake depth of 2161 ft (658.7 m), the initial hydrostatic pressure of the formation is 7.66 MPa. The first flow period lasted 16.6 min and was conducted with the flowing bottom-hole pressure (FBHP) less than the estimated hydrate stability pressure (Fig. 4) at the flowing bottom-hole temperature (FBHT). The hydrate stability curves inFig. 4 were calculated using the Moridis (2003) correlation given the downhole measured temperature both at the onset of the flow test and at the temperature measured throughout the flow test. The subsequent build-up appeared to include non-porous-media effects such as a slow pressure increase and an inflection in the pressure curve near the 1.4-h point of the test. The first build-up was allowed to last 52.9 min and reached only 6.1 MPa before the second flow period began. This dampened pressure build-up was characteristic of the two C-unit MDT tests and was explored further in the history-matching effort focused on the C-2 MDT test.

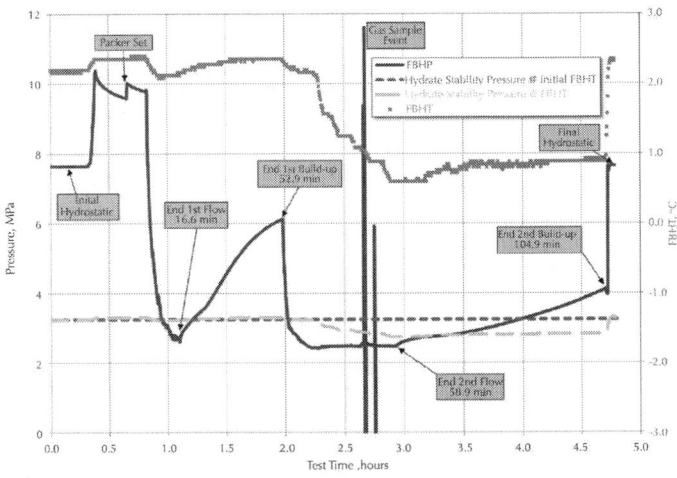

Figure 4: Downhole measured flowing bottom-hole pressure (FBHP) and temperature for the C1 MDT experiment. Yellow trace indicates predicted (Moridis, 2003) gas-hydrate stability pressure at measured temperature.

The second flow period of the C-1 test was again conducted with the FBHP less than the hydrate stability pressure. During this longer (58.9 min) flow period a gas sample was taken at the 2.67-h mark into the C-1 test (shown in Fig. 4). The pressure build-up following the second flow period was severely dampened and was ended after 104.9 min.

The C2 MDT Flow Test

The design of the C2 MDT flow test (Fig. 5) differed from the C1 test in that during the first flow period the well pressure was kept above the in situ hydrate dissociation pressure (i.e., the well pressure remains above the gas hydrate equilibrium pressure based on the in situ temperature and the FBHT) as shown inFig. 5. As a result, the only methane extracted from the reservoir during this period was the very small amount that was dissolved in the extracted formation water (i.e., no free gas was detected at the MDT intake port during this first drawdown period). The best estimate of the produced water and gas volumes are shown in Fig. 6.

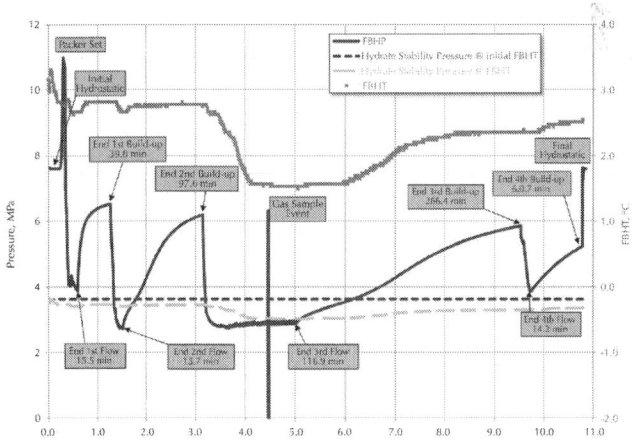

Figure 5: Downhole measured flowing bottom-hole pressure (FBHP) and temperature for the C2 MDT experiment. Yellow trace indicates predicted (Moridis, 2003) gas-hydrate stability pressure at measured temperature.

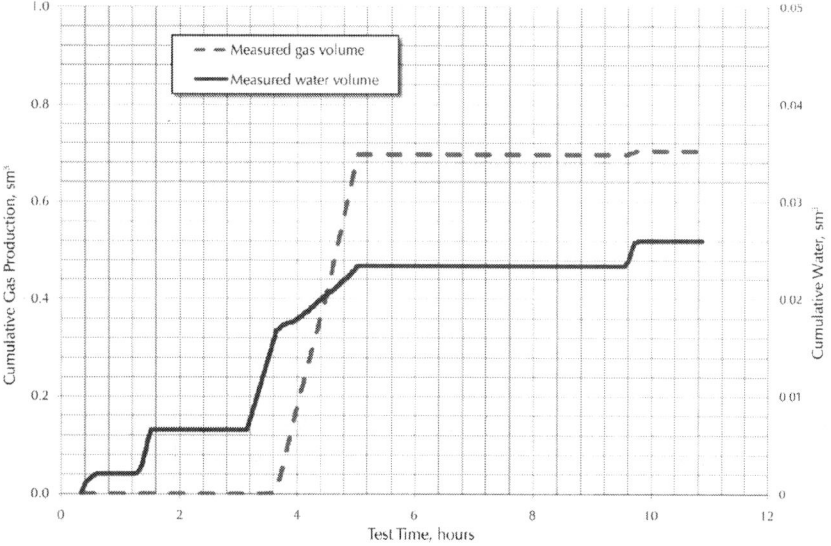

Figure 6: Estimated volumes of gas and water pumped from the test zone during the C2 MDT flow test.

By analyzing the pressure response of the reservoir after the first flow period, an estimate of the effective permeability of the formation in the presence of hydrate can therefore be obtained. The importance of this parameter cannot be overstated as it is one of the key parameters controlling the potential productivity of any reservoir.

During the second and third flow periods the pressure was reduced below the expected gas hydrate equilibrium pressure, thereby resulting in dissociation of gas hydrate and the release of free gas into the formation (see Fig. 5). The optical analyzer indicated that during the second pressure drawdown period no to little gaseous methane was pumped through the MDT tool, which initially was in contrast to the expectation of gas production with hydrate dissociation. Evidence of produced gas, however, was indicated during the pressure build-up response to the second pressure drawdown. The pressure build-up response after the first pressure drawdown was characteristic of the recovery in a confined aquifer. The prolonged pressure recovery after the second pressure drawdown indicated compressible gas in the annular space of the MDT above the screened inlet.

During the third (and longer) flow period, the pressure was once again reduced to a point below the hydrate equilibrium pressure, this time over a sufficiently longer period that resulted in the measurable production of both formation water and methane gas. The pressure recovery after this flow period was even more prolonged than that after the second. Both the second and third pressure–recovery curves display an inflection point in the experimentally observed pressure (see Fig. 5), potentially indicating some type of flow regime transition or other significant change in the physical processes influencing the pressure build-ups.

The D1 MDT Flow Test

There were three different flow periods conducted in the D1 MDT flow test. The first flow (11.1 min) and the extended second flow (229.4 min) periods were conducted with the FBHP greater than the hydrate stability pressure. In these two flow periods the hydrate remained undisturbed. This result is evidenced by the pressure build-up curves following the flow periods (Fig. 7) having what would be considered characteristic of a porous-media response. The third flow period of the D1 MDT test was conducted with the FBPH less than the hydrate stability pressure resulting in hydrate dissociation and gas production (Fig. 7). A gas sample was obtained from this third flow period. Throughout the D1 test decreasing pump performance due to extended pumping times and wear due to fine sediments in the flow stream was detected. The third build-up ended prematurely due to a packer seal failure and the system abruptly returned to in situ hydrostatic conditions.

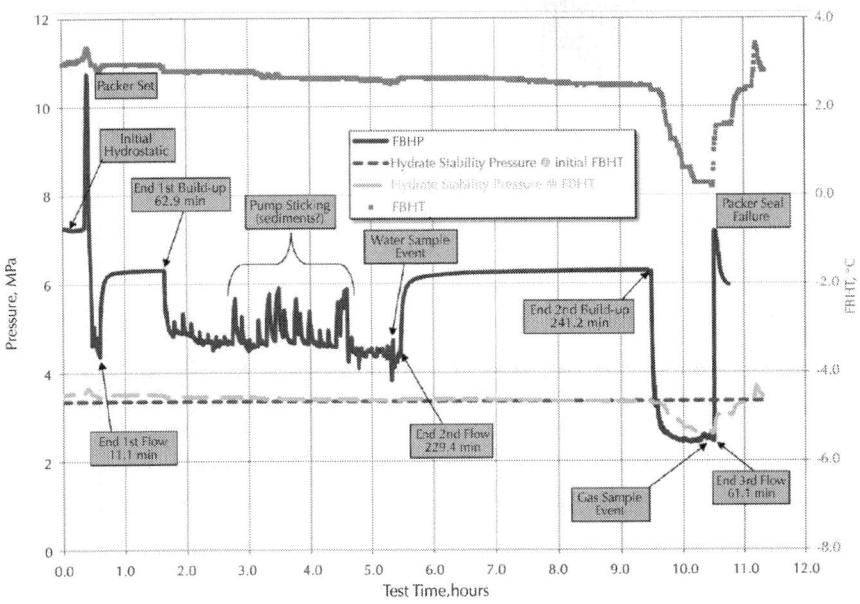

Figure 7: Downhole measured flowing bottom-hole pressure (FBHP) and temperature for the D1 MDT experiment. Yellow trace indicates predicted (Moridis, 2003) gas-hydrate stability pressure at measured temperature.

The D2 MDT Flow Test

Two different flow periods (Fig. 8) were conducted in the D2 MDT flow test. The first flow (12.4 min) maintained the pressure greater than the hydrate stability pressure (Fig. 8) and again a classic porous-media response was observed on the first build-up. The ability to repeatedly maintain the FBHP above the hydrate stability pressure and observe classic porous-media responses allow us to characterize the initial permeability of the hydrate reservoir. As discussed in the history-matching exercise later, the determination of this initial permeability is a significant result of this effort.

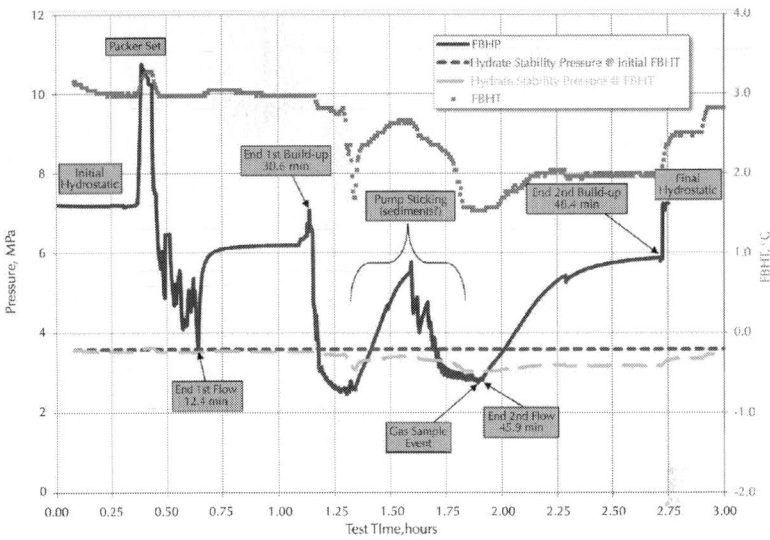

Figure 8: Downhole measured flowing bottom-hole pressure (FBHP) and temperature for the D2 MDT experiment. Yellow trace indicates predicted (Moridis, 2003) gas-hydrate stability pressure at measured temperature.

The second flow period of the D2 flow test was conducted at a pressure less than the hydrate stability pressure; however, pump wear impeded the ability to compress and therefore cut the flow test short. Nevertheless, a gas sample was obtained from the hydrate dissociation during the second D2 flow period and the dampened pressure build-up was again observed.

MDT Test Results and Analysis

After reviewing the raw rate and pressure responses from the four test sequences, the C2 sequence was selected as having the highest likelihood of providing interpretable trends using conventional analytical pressure transient methods. The sequence of three drawdown then build-up sequences was loaded into a proprietary pressure transient program and synchronized with rate profiles generally estimated from the MDT pump stroke data. Each of the build-up and drawdown periods was analyzed using classic straight line techniques as well as log–log type curve fitting methods (Earlougher, 1977 and Gringarten, 1987).

As is typical, the drawdown trends were distorted by the inability to maintain a perfectly constant flowrate at each flow period. After 0.1 h, storage effects had dissipated, but each pump stroke imparted a distinct pressure pulse into the overall response, resulting in a noise level that precluded definitive parameter estimates (Fig. 9).

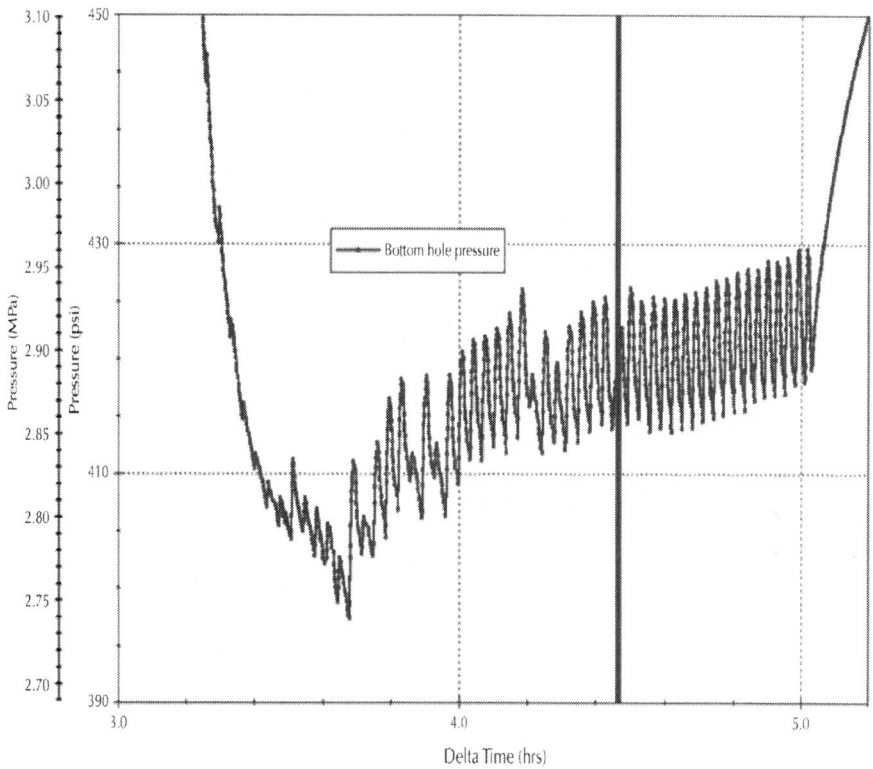

Figure 9: Detailed pressure response of the C2 MDT experiment showing the pressure pulses due to individual pump strokes.

Attention then turned to the first build-up period where a classic porous-media response was observed. Wellbore storage and permeability estimates were performed using the Ryder Scott proprietary pressure transient analysis tool, PTA (Fig. 10). The estimated permeability of the hydrate-bearing media was found to be approximately 0.5 mD using this pressure transient analysis. This estimate will be further analyzed through the history-matching efforts discussed later in the paper. Although the flow period was only 15 min, a clean and interpretable build-up response was seen following the cessation of pumping. A small wellbore storage constant (5.21E-6 bbl/psi) was calculated from an extrapolation of the initial trend on the linear delta time versus pressure plot (Fig. 11) again using the Ryder Scott PTA tool.

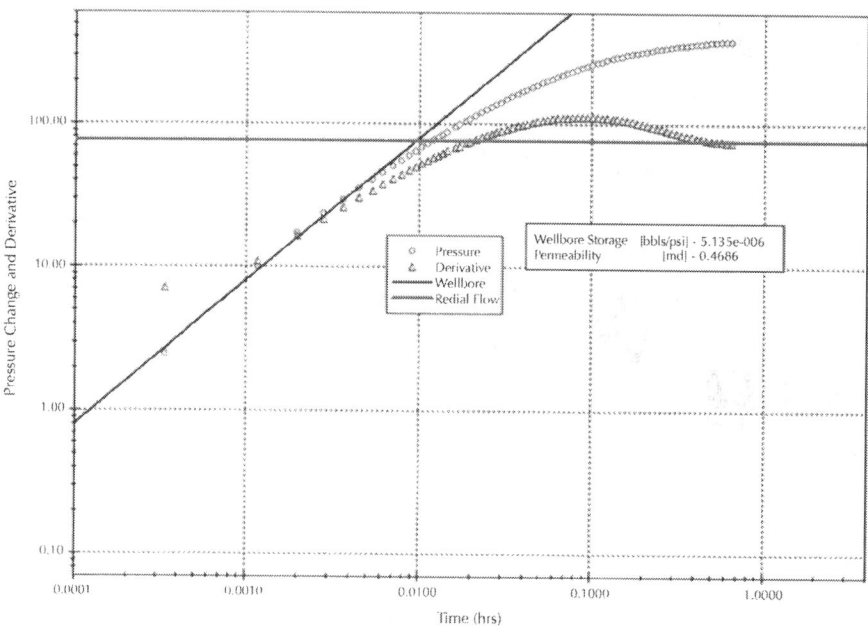

Figure 10: The log–log pressure response of the 1st build-up of the C2 MDT experiment showing classical pressure build-up response.

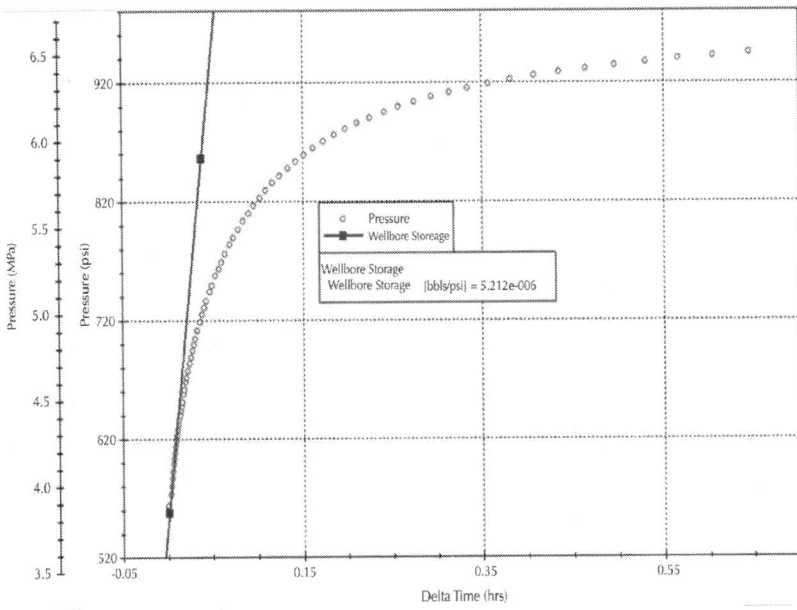

Figure 11: Linear time versus pressure plot for the first flow period of the C2 MDT test that is used to give a first estimate of very early time wellbore storage constant.

Using the very short first build-up test, a rough estimate of the permeability of the hydrate-filled sediment to water is 0.5 mD as shown in Fig. 10. This value is in line with prior estimates (Mallik) of water permeability in hydrate-saturated rock (Kurihara et al., 2008). More detailed analyses using both gas and water flow capacity indicate permeability to be closer to 0.2 mD as discussed later in the history-matching discussion. Given that the intrinsic permeability of the dry rock is on the order of 0.5–2 darcies (Winters et al., 2011), this represents a fairly consistent and predictable outcome of severe permeability reduction as hydrate saturation increases.

The second and third flow periods, which include gas hydrate dissociation effects, were not as useful for determining reservoir petrophysical parameters. Fig. 12 and Fig. 13 show the log–log character of the 2nd and 3rd flow periods, respectively. Like the raw pressure data shown in Fig. 5, the build-up during these periods was "delayed" and repressed compared to what was seen in the response to the first depressurization period. In fact, the 3rd period showed a more severe departure from the normal behavior, indicated by the inflection point near one hour, than the second. Notice that the first build-up period, the period in which no hydrate was dissociated and no gas produced, did not result in an inflection point in the pressure derivative curve shown in Fig. 10. Fig. 14 and Fig. 15 show wellbore storage trends for the second and third periods, in which each show an increase in the wellbore storage constant over the previous build-up period.

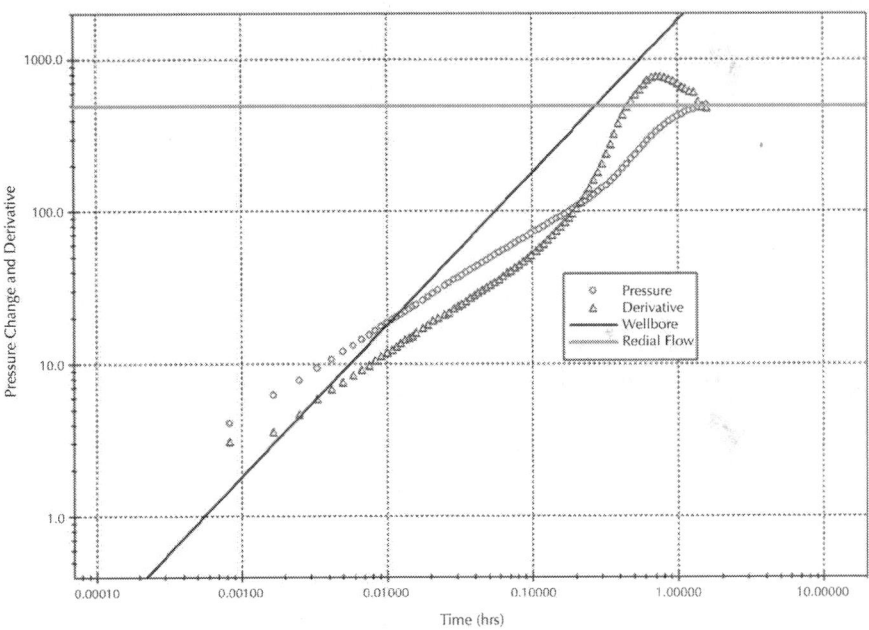

Figure 12: The log–log plot for Period 2 of the C2 MDT test that should be similar to period 1 (Fig. 10). Note the distinctive inflection point that occurs in the pressure derivative curve at roughly 0.2 h (~12 min).

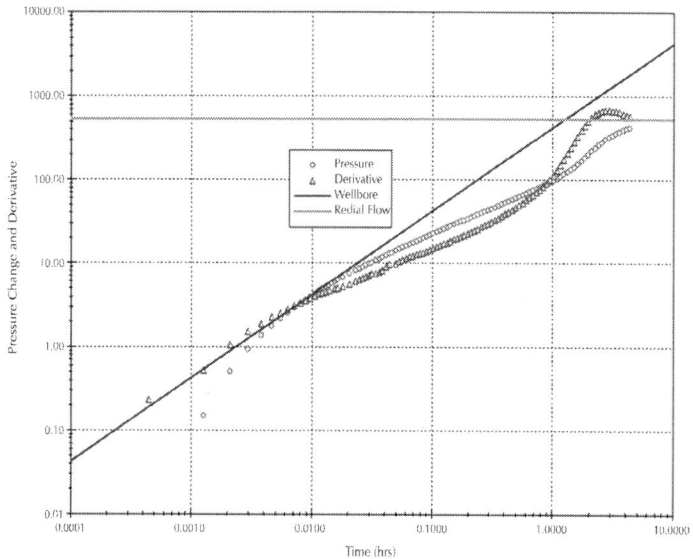

Figure 13: The log–log plot for Period 3 of the C2 MDT test that should be similar to Period 1 (Fig. 10) but also has the same inflection point anomaly as Period 2 (Fig. 12). In this period, the break in slope occurs at 1 h instead of 0.2 h as in the prior period.

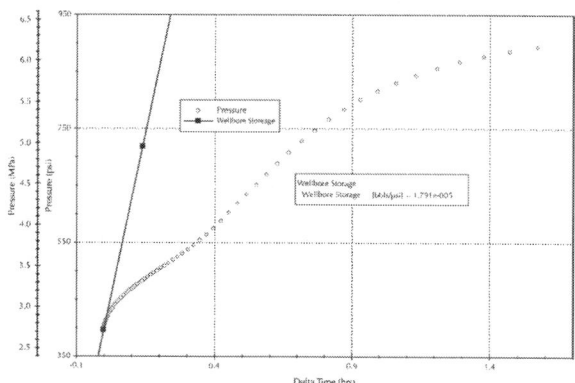

Figure 14: Linear time versus pressure plot for the second flow period of the C2 MDT test that is used to give a first estimate of very early time wellbore storage constant. Compared to Period 1, the wellbore storage constant is larger.

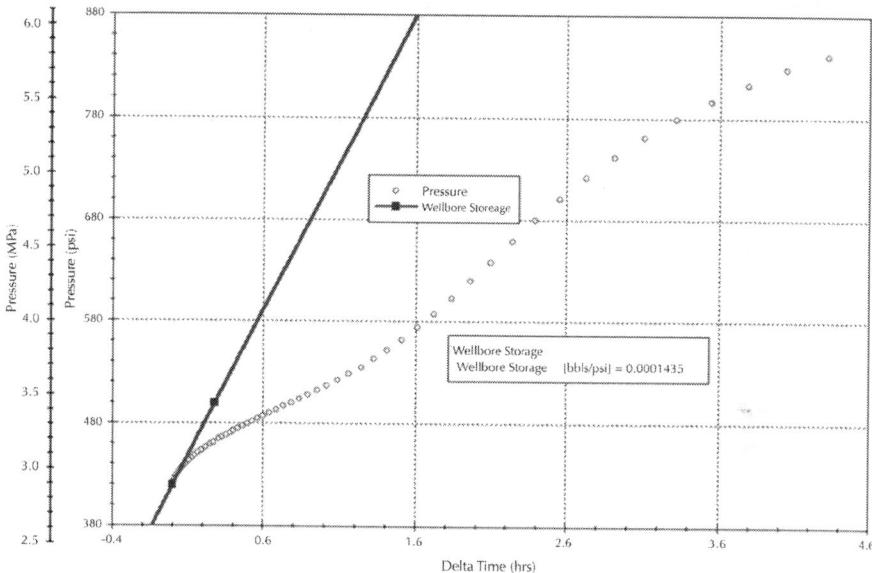

Figure 15: Linear time versus pressure plot for the third flow period of the C2 MDT test. Compared to periods 1 and 2 and shown in Fig. 16, the wellbore storage constant is larger yet.

A plot of cumulative volume produced versus wellbore storage constant (Fig. 16) shows a consistent trend toward higher compressibility as each new flow period ended, indicating an increase in gas filled volume in the annular space between the tool and the wellbore. This changing wellbore storage effect was repeatable and also was evident in other MDT build-ups, with the build-up responses during the C1 test showing an even more pronounced changing wellbore storage effect.

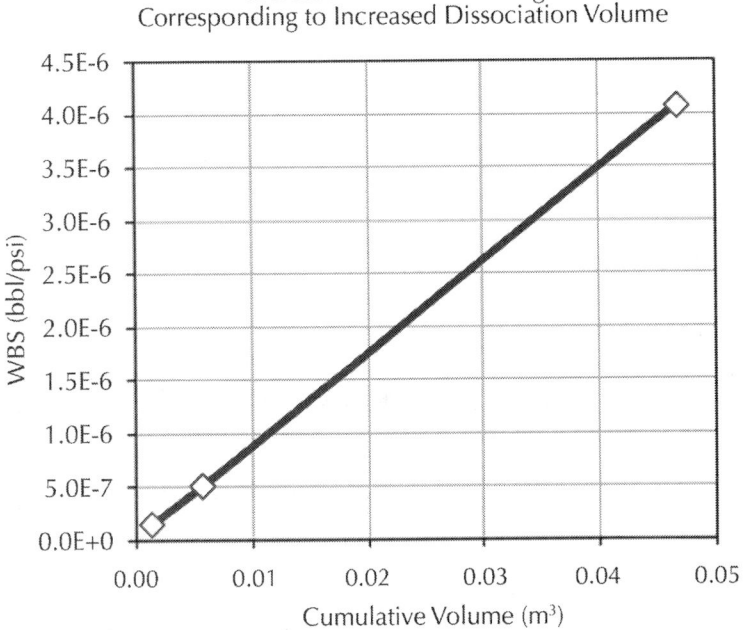

Figure 16: Wellbore storage constant versus estimated cumulative volume produced. Although this shows a linear trend, it is reasonable to assume that once the entire void space between the exit port and the wellbore annulus fills with gas, a constant wellbore storage constant will emerge.

Fortunately, this simple explanation can account for the much of the deviation from normal tight gas reservoir behavior in the pressure build-up responses. Reservoir pressure and native water permeability were interpretable from these measurements, but further detail was obscured by the mechanical issues related to changing wellbore storage. As will be described in the code comparison discussion below, this annular void space became the key parameter needed to understand the MDT pressure responses.

Ironically, the C2 MDT test was selected to be interpreted because it showed the most "classic" and interpretable response with no packer failure or overpull at the end of the test. Meanwhile, other tests showed both plugging by formation sand production (D1 and D2) and packer failure (D1) indicating strong dissociation and wellbore sloughing; which are characteristics of a more productive zone than that tested in the C2 test. Just as oil well exploration in the early 1900's depended on blowouts to identify highly productive zones, "failed" MDT tests may eventually be shown to correlate to high productivity hydrates.

Comparative Analysis of C2 Test Results

The National Energy Technology Laboratory (NETL) and the U.S. Geological Survey (USGS) are guiding a collaborative, international effort to compare methane hydrate reservoir simulators. The intentions of the effort are: (1) to exchange information regarding gas hydrate dissociation and physical properties enabling improvements in methane hydrate reservoir modeling, (2) to build confidence in all the leading simulators through exchange of ideas and cross-validation of simulator results on common datasets of escalating complexity, and (3) to establish a depository of gas hydrate related experiment/production scenarios with the associated predictions of these established simulators that can be used for ongoing and future comparison purposes. To achieve these goals, a team of researchers was brought together to construct a series of problems designed to test/compare the performance of the leading gas hydrate simulators. Participating in this effort are researchers utilizing five distinct simulators. These simulators are: CMG-STARS (Computer Modeling Group Ltd, 2007), MH21-HYDRES (MH-21 Research Consortium), STOMP-HYDRATE (White and Oostrom, 2006), TOUGH + HYDRATE (Moridis et al., 2005b), and HydrateResSim (Moridis et al., 2005a). To date this team has constructed a series of seven problem sets. The first five of these problem sets examined various facets of the multiphase flow/equilibrium behavior necessary to model this complex system accurately, and are reported on elsewhere (Wilder et al., 2008). In this work we will present results of Problem 6, and describe preliminary results related to the seventh problem set.

The objective of the sixth problem, described in this work, considered by the Methane Hydrate Reservoir Simulator Code

Comparison Study was to analyze the data obtained from an actual hydrate well test. The data utilized were obtained from the Mount Elbert well which was drilled as part of the cooperative DOE-BPXA research project.

For modeling purposes, the first through the third flow and recovery periods of the second MDT experiment (the C2 test) conducted on the Mount Elbert well were selected. The MDT experiments involved alternating flow periods (of various durations), using a positive-displacement pump, and build-up phases, during which there was no pumping.

HISTORY MATCHING SETUP AND RESULTS

In an effort to conduct this phase of the Methane Hydrate Reservoir Simulator Code Comparison Study in a manner that would reflect how actual history matches would be conducted using the codes separately, each modeler was given the freedom to determine the approach to conducting these history matches with respect to determination of the numerical grid, approach to finding fitting parameters, etc. The only constraints placed on the efforts were based on the experimental setup (i.e., the location of the tool in the formation, the size of the wellbore, etc), experimentally observed properties of the formation (porosity, initial saturations, temperatures, etc), and the MDT test data.

The experimental data discussed above was utilized in this study by incorporating it into the numerical models used to construct the desired history matches. First, a schematic of the well and the placement of the MDT tool in the wellbore was constructed (Fig. 17) based on the model setup. A two-dimensional cylindrical grid was used to model the annular space in the well and the hydrate-bearing formation that extended radially outward from the wellbore.

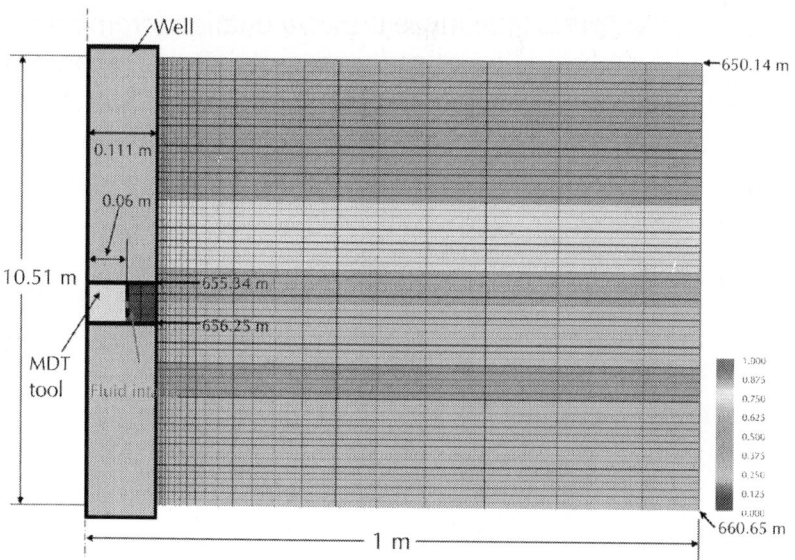

Figure 17: Schematic showing the relation of the MDT tool during the C2 test to the 10-m thick C unit. Also shown is an example of the reservoir (and annular space) gridding used in the simulations. The color scale indicates the gas-hydrate saturation. From Kurihara et al. (2011).

Simulations

Five simulators (CMG-STARS, STOMP-HYDRATE, TOUGH + HYDRATE, MH21-HYDRES, and HydrateResSim) were independently used to conduct history matches based on the experimental data collected during the three flow periods. During the simulations being reported on here, the models used the observed pumping (flow) periods as specified boundary conditions (ie., the simulated pressure at the location of the MDT inlet was set to the experimentally observed pressure during the flow periods). Model parameters were then adjusted to obtain the best possible fits to the observed temperature and produced fluid volumes, as well as the pressure during the pressure build-up periods (ie., after the cessation of each pumping event). Based on the nature of the data obtained from the MDT experiments, it was decided that the most accurate data were the pressures reported by the tool, followed by the temperature and produced fluid volumes. The latter two were of a lower quality for the following reasons: the temperature was felt to be of reduced accuracy due to the location of the sensor and the possibility that it was at various times in thermal equilibrium with formation water and/or free gas, and that the temperature might not necessarily accurately reflect the instantaneous (average) temperature of the formation at the physical location of the tool inlet, rather it was measuring the temperature of the fluid in contact with the tool.

With respect to the produced fluid volumes, the uncertainties were related to the necessity of having to interpret the volume of each fluid produced as a result of each pump stroke based on the pressure response of the pump chamber to the compressibility of the fluid(s) in the chamber during any particular stroke. It was therefore felt that the produced volumes contained the greatest error, and the pressure the least. As a result, in constructing the history matches, the observed pressure during the build-up periods was used as the primary fitting criteria, with temperature and produced fluid volumes given secondary importance.

The final history matches obtained by the various groups running the simulators are summarized in Fig. 18. General conclusions concerning these results are discussed in the next section. The investigators used a wide range of approaches in constructing their individual history matches. For example, the number of total grid cells used to represent the modeled portion of the formation ranged from 360 to over 10,000. Some investigations included the solubility of methane in water as well as the formation water's observed salinity, while others ignored both. In all of the cases very good fits were obtained with respect to the observed pressure during the various build-up phases, however in none of them was a reasonable match to the estimated volume of produced gas obtained. This was expected due to the uncertainty in the produced gas rates, and because the gas rates were not matched as part of the history-matching exercise. General comments concerning these results are discussed in the next section.

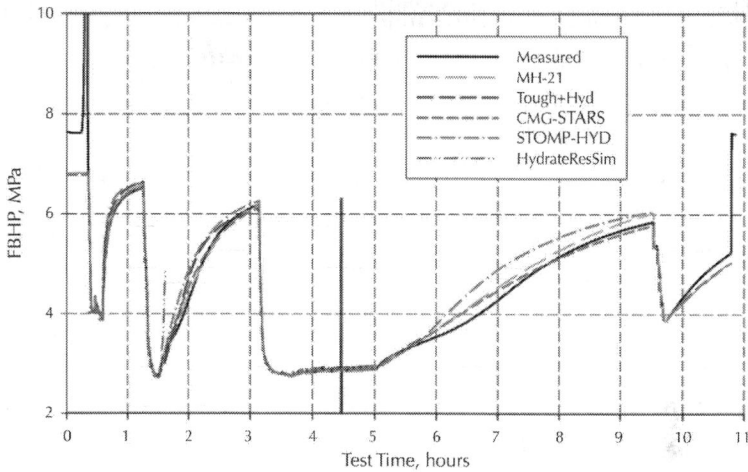

Figure 18: Summary plot of the history match to the C2 MDT test.

DISCUSSION

History Match – First Pressure Build-up

As discussed previously, during the first flow period the reservoir pressure never went below the hydrate equilibrium pressure. As a result, no hydrate dissociated, and the recorded data can therefore be used to reconstruct an initial estimate of the formation permeability (in the presence of hydrate) in the vicinity of the MDT tool. Based on the history match results of the various simulators, the effective permeability of the formation is estimated to be in the range of 0.12–0.17 mD (with an intrinsic (no gas hydrate present) or absolute permeability of approximately 1 D).

It is noteworthy that even given all of the differences between the approaches utilized with the four different simulators, all of the history matches from the various models to this portion of the pressure data resulted in an estimate of the effective permeability in the same range (0.12–0.17 mD). Though this estimate may only be reflective of the reservoir in the very near vicinity of the borehole, it represents perhaps the best information to date on this key parameter.

History Match – Second and Third Pressure Build-ups

Initial attempts to construct history matches using the second and third flow/build-up periods were not very successful. This difficulty was overcome when an annular space was explicitly included around the MDT tool which accounted for wellbore storage of reservoir fluids. After the inclusion of this annular space, very good pressure matches were readily obtained. Based on the results from the various simulations, it seems that fluid segregation in this annular space plays a key role in the general shape of the recovery curves. Without this space, the simulated recovery curves have the more traditional shape seen during the first build-up phase (prior to the release of any gas from hydrate in the formation).

As was also mentioned above, an appreciable amount of gas was not produced during the second flow period, yet all of the simulators indicate that an appreciable amount of hydrate did dissociate, and a corresponding amount of free gas was released into the formation during this time. With the annular space included in the numerical simulations, it was observed that as gas migrated into the region near the MDT tool inlet, fluid segregation resulted in the accumulation of free gas in the region above the inlet, resulting in the production of only formation water during the second flow period. Only after sufficient gas had migrated to this region (some time during the third flow period) and the water level had decreased below the tool inlet did appreciable amounts of free gas begin to be produced. While inclusion of an annular space did allow the good history matches to be achieved (with respect to the pressure), there is one drawback to including this effect. Due to the small amount of fluid produced during the experiment, segregated fluid flow in the annular space had a significant impact on the observed pressure build-ups. Unfortunately, none of the codes under consideration include the physical/mathematical models necessary to rigorously model instantaneous fluid segregation, in a fluid-filled annular space. As a result, there is a possibility that the model parameters determined during the history matches may have been skewed by the inclusion conditions where a phenomenon the models were not specifically designed to simulate was important to the results. Since parameters would be useful as a starting point of a detailed sensitivity analysis directed at assessing potential production from such a formation, they should not be interpreted as "the" parameters from which a single prediction of the potential productivity of the formation should be made.

Another interesting characteristic of the pressure build-ups is that the latter two evidenced an inflection point (for example, examine the blue trace in Fig. 5 shortly after a time of 6 h). The change in curvature of the build-up at this point may be indicative of a change in the character of the fluid flow in the formation. Such a change may be due to, flow regime transition (perhaps involving the segregated fluid flow in the annular space), effects of hydrate reformation (or lack thereof) on the migration of fluids toward the MDT tool, or disappearance of free gas in the formation. Because the simulators do not explicitly include models for segregated flow in an annular space, we are unable to attribute this transition to a particular phenomenon.

Experimental Simulation of the C2 MDT Data

A simple physical experiment has been performed at the Colorado School of Mines (CSM) to determine if the increasing gas volume in the annular space in the wellbore next to the tested zone could have caused the anomalous build-up trends seen during the C2 test sequence. The experimental setup, shown in Fig. 19, consisted of a Plexiglas® cylinder (vessel) connected to a constant-pressure water and gas source with precision needle valves regulating the flow into the vessel. The constant-pressure water and gas source represented an infinite reservoir with two needle valves modeling a fixed permeability medium (reservoir near the wellbore) connected to gas and water sources. Adjusting the pressure drop across these valves changed the rate of influx of water and gas into the system. A third valve represented the offtake point in the Schlumberger MDT tool. It was used to remove fluid from the center of the vessel during the "flow" periods.

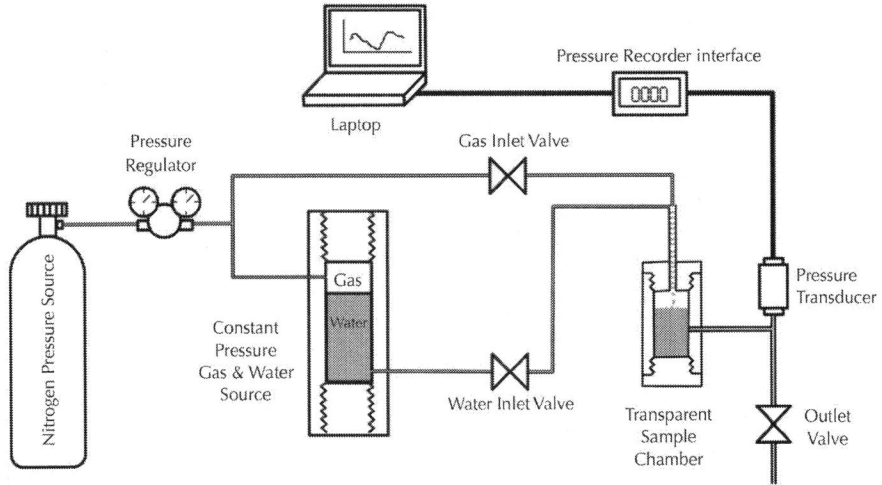

Figure 19: Schematic of the experimental setup.

This experiment demonstrated that the pressure data collected could be replicated with laboratory procedures independent of hydrate formation or dissociation (Fig. 20).

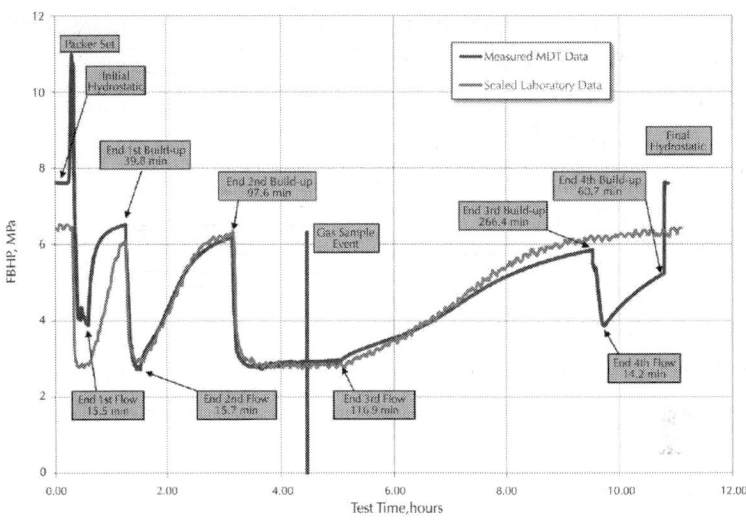

Figure 20: Experimental reproduction of the C2 MDT pressure-response data.

Initially, the pressure was equilibrated throughout with the water-filled vessel, with the source valves slightly open and the offtake valve closed. Once the vessel reached "reservoir" pressure, the offtake valve was opened to allow a small amount of fluid to flow from the vessel. The vessel pressure decreased rapidly, and then began to stabilize as flow from the two source valves re-charged the vessel. After the short flow period, the offtake valve was closed and the pressure was allowed to build up once again. As seen in Fig. 20, the first build-up reached stable pressure relatively quickly because little gas was present in the system.

Without changing the pressure settings on the source valves, the offtake valve was again opened for the second, longer, flow period. This flow period allowed more gas to build up in the vessel. When the offtake valve was closed, the second build-up took longer than the first because of the presence of the compressible gas. This effect can be seen in Fig. 20 as the second build-up. Once the build-up pressure stabilized again, the offtake valve was opened and remained open for a much longer period of time, allowing a large amount of gas to accumulate in the system. As can be seen in Fig.

20, the final (3rd) build-up took much longer, much like the MDT data from the Mount Elbert Well.

For the entire experiment, the settings on the gas and water valves remained unchanged, so the rate of gas and liquid into the system was likely a good proxy for Darcy flow into the void space around the MDT tool. The flow and build-up period lengths were proportional to the times used in the reservoir tests and resulted in an excellent proportional pressure response, as illustrated in Fig. 20. Applying Occam's razor ("the simplest explanation is usually correct") the anomalous pressure build-up response in the C2 MDT Pressure response was likely caused by varying gas volumes and changing wellbore storage effects.

CONCLUSIONS

Independent analysis of the MDT data from the Mount Elbert well utilizing five simulators (CMG-STARS, STOMP-HYDRATE, TOUGH + HYDRATE, MH21-HYDRES, and HydrateResSim) has led to very important insights into the potential behavior of hydrate-bearing formations such as the one at Mount Elbert. Through the history-matching efforts outlined in this paper, it was determined that the annular space in the MDT tool provided sufficient wellbore storage to significantly affect the nature of the pressure build-up curves. Once the models adequately included the annular space of the MDT tool, all of the participating reservoir simulators matched the pressure recovery curves of the MDT experiments. Additionally, the initial depressurization period of the C2 test proved invaluable in allowing for the determination of the initial permeability of the hydrate reservoir.

The parameters determined as part of the history match being reported on here should be viewed as informative, but not definitive. Because of the limited extent to which the formation as a whole was sampled by this test, and because there is an as yet unknown impact of having to include the annular space (due to the small volume of fluids produced during the test), there is insufficient evidence on which to base an assertion that the parameters being reported here would be representative of the formation in general. However, these parameters (representing the best "local" estimates available to date) would be extremely useful as a starting point for a detailed sensitivity analysis directed at assessing potential production from such a formation.

As the community moves forward with subsequent *in situ* test of gas hydrate formations the choice of downhole test need to be considered carefully. MDT data may be useful in estimating local permeabilities; "global" (or "average") permeability estimates would require flow tests that sampled a much larger portion of the reservoir than is possible with the MDT tool. To understand why such long tests are so important in the case of hydrates, one should consider that during hydrate dissociation/formation the pore space available for fluid flow changes (due to hydrate dissociation and/ or reformation), thereby impacting the apparent permeability of the formation. Thus, a short-term test is not indicative of the fully developed flow/behavior of the formation after significant hydrate has dissociated/reformed. Exactly how long such a test would need to be in order to provide optimum data is an open (and very interesting) question.

ACKNOWLEDGEMENTS

The authors would like to thank the National Energy Technology Laboratory of the U.S. Department of Energy, the U.S. Geological Survey, the Japan MH-21 project, and BP Exploration (Alaska) for supporting this effort. We would also like to acknowledge the Mount Elbert science party for sharing the data obtained at Mount Elbert for use in our history-matching and production simulations. Finally, the authors would like to thank Michael Batzle for his supervision of the experimental simulation and Marisa Rydzy for the original experimental apparatus.

REFERENCES

1. Bily, C., Dick, J.W.L., 1974. Naturally occurring gas hydrates in the Mackenzie Delta, NWT. Bulletin of Canadian Petroleum Geology 22 (3), 340e352.

2. Boswell, R., Hunter, R., Collett, T., Digert, S., Hancock, S., Weeks, M., Mount Elbert Science Team, 2008. Investigation of gas hydrate-bearing sandstone reservoirs at the "Mount Elbert" stratigraphic test well, Milne Point, Alaska. In: Proceedings of the 6th International Conference on Gas Hydrates, Vancouver, British Columbia, Canada.

3. Computer Modeling Group Ltd, 2007. CMG STARS. Computer Modeling Group Ltd., Calgary, Alberta, Canada.

4. Dallimore, S.R., Collett, T.S. (Eds.), 2005. Scientific Results from the Mallik 2002 Gas Hydrate Production Research Well Program, Mackenzie Delta, Northwest Territories, Canada.

5. Earlougher Jr., R.C., 1977. Advances in Well Test Analysis. In: Monograph Series. SPE, Richardson, TX.

6. Gringarten, A.C., 1987. Type-curve analysis: what it can and cannot do. Journal of Petroleum Technology. 39 (1), 11e14.

7. Hunter, R.B., Collett, T.S., Boswell, R.M., Anderson, B.J., Digert, S.A., Pospisil, G., Baker, R.C., Weeks, L.M., 2011. Mount Elbert Gas Hydrate Stratigraphic Test Well, Alaska North Slope: overview of scientific and technical program. Journal of Marine and Petroleum Geology 28 (2), 295e310.

8. Inks, T., Lee, M., Agena, W., Taylor, D., Collett, T., Hunter, R., Zyrianova, M., 2009. Seismic prospecting for gas hydrate and associated free-gas prospects in the Milne Point Area of Northern Alaska. In: Collett, T., Johnson, A., Knapp, C., Boswell, R. (Eds.), Natural Gas Hydrates: Energy Resource and Associated Geologic Hazards. American Association of Petroleum Geologists Memoir 89.

9. Kurihara, M., Funatsu, K., Ouchi, H., 2008. Analysis of the JOGMEC/NRCAN/Aurora Mallik gas hydrate production test through numerical simulation. In: Proceedings of the 6th

International Conference on Gas Hydrates, Vancouver, British Columbia, Canada.

10. Kurihara, M., Sato, A., Funatsu, K., Ouchi, H., Masuda, Y., Narita, H., Collett, T.S., 2011. Analysis of formation pressure test results in the Mount Elbert methane hydrate reservoir through numerical simulation. Journal of Marine and Petroleum Geology 28 (2), 502e516.

11. MH-21 Research Consortium. http://www.mh21japan.gr.jp/english.

12. Moridis, G.J., 2003. Numerical studies of gas production from methane hydrates. Society of Petroleum Engineers Journal 32 (8), 359e370.

13. Moridis, G.J., Kowalsky, M.B., Pruess, K., 2005a. HydrateResSim Users Manual: A Numerical Simulator for Modeling the Behavior of Hydrates in Geologic Media. Contract No. DE-AC03-76SF00098. Department of Energy, Lawrence Berkeley National Laboratory, Berkeley, CA.

14. Moridis, G.J., Kowalsky, M.B., Pruess, K., 2005b. TOUGH-Fx/HYDRATE v1. 0 User's Manual: A Code for the Simulation of System Behavior in Hydrate-bearing Geologic Media. Report LBNL-58950. Lawrence Berkeley National Laboratory, Berkeley, CA.

15. Rose, K.R., Boswell, R.M., Collett, T.S., 2011. Mount Elbert Gas Hydrate Stratigraphic Test Well, Alaska North Slope: Coring operations, core sedimentology, and lithostratigraphy. Journal of Marine and Petroleum Geology 28 (2), 311e331.

16. Torres, M.E., Collett, T.S., Rose, K.K., Sample, J.C., Agena, W.F., Rosenbaum, E.J., 2011. Pore fluid geochemistry from the Mount Elbert Gas Hydrate Stratigraphic Test Well, Alaska North Slope. Journal of Marine and Petroleum Geology 28 (2), 332e342.

17. White, M.D., Oostrom, M., 2006. STOMP Subsurface Transport Over Multiple Phase: User's Guide PNNL-15782 (UC-2010). Pacific Northwest National Laboratory, Richland, Washington.

18. Wilder, J., Moridis, G., Wilson, S., Kurihara, M., White, M., Masuda, Y., Anderson, B., Collett, T., Hunter, R., Narita, H., Pooladi-Darvish, M., Rose, K., Boswell, R., 2008. An international effort to compare gas hydrate reservoir simulators. In Proceedings of the 6th International Conference on Gas Hydrates, Vancouver, British Columbia, Canada.

19. Winters, W.J., Walker, M.M., Hunter, R.B., Collett, T.S., Boswell, R.M., Rose, K.K., Waite, W.F., Torres, M.E., Patil, S.L., Dandekar, A.Y., 2011. Physical properties of sediment from the Mount Elbert Gas Hydrate Stratigraphic Test Well, Alaska North Slope. Journal of Marine and Petroleum Geology 28 (2), 361e380.

5

Latest Development on Membrane Fabrication for Natural Gas Purification: A Review

Dzeti Farhah Mohshim, Hilmi bin Mukhtar, Zakaria Man, and Rizwan Nasir

Chemical Engineering Department, Universiti Teknologi Petronas, Bandar Seri Iskandar, Perak Darul Ridzuan, 31750 Tronoh, Malaysia

ABSTRACT

In the last few decades, membrane technology has been a great attention for gas separation technology especially for natural gas sweetening. The intrinsic character of membranes makes them fit for process escalation, and this versatility could be the significant factor to induce membrane technology in most gas separation areas. Membranes were synthesized with various materials which depended on the applications. The fabrication of polymeric membrane was one of the fastest growing fields of membrane technology. However, polymeric membranes could not meet the separation performances required especially in high operating pressure due to deficiencies problem. The chemistry and structure of support materials like inorganic membranes were also one of the focus areas when inorganic membranes showed some positive results towards gas separation. However, the materials are somewhat lacking to meet the separation performance requirement. Mixed matrix membrane (MMM) which is comprising polymeric and inorganic membranes presents an interesting approach for enhancing the separation performance. Nevertheless, MMM is yet to be commercialized as the material combinations are still in the research stage. This paper highlights the potential promising areas of research in gas separation by taking into account the material selections and the addition of a third component for conventional MMM.

INTRODUCTION

Natural gas can be considered as the largest fuel source required after the oil and coal [1]. Nowadays, the consumption of natural gas is not only limited to the industry, but natural gas is also extensively consumed by the power generation and transportation sector [2]. These phenomena supported the idea of going towards sustainability and green technology as the natural gas is claimed to generate less-toxic gases like carbon dioxide (CO_2) and nitrogen oxides (NO_x) upon combustion as shown in Table 1 [3].

Table 1: Fossil fuel emission levels (pounds per billion Btu of energy input)

Fuel sources/pollutant(pound/ BTU)	Natural gas	Oil	Coal
Carbon dioxide	117,000	164,000	208,000
Carbon monoxide	40	33	208
Nitrogen oxides	92	448	457
Sulphur dioxide	1	1,122	2,591
Particulates	7	84	2,744
Mercury	0.000	0.007	0.016

However, pure natural gas from the wellhead cannot directly be used as it contains undesirable impurities such as carbon dioxide (CO_2) and hydrogen sulphide (H_2S) [4]. All of these unwanted substances must be removed as these toxic gases could corrode the pipeline since CO_2 is highly acidic in the presence of water. Furthermore, the existence of CO_2 may waste the pipeline capacity and reduce the energy content of natural gas which eventually lowers the calorific value of natural gas [5].

Conventionally, natural gas treatment was predominated with some methods such as absorption, adsorption, and cryogenic distillation. But these methods require high treatment cost due to regeneration process, large equipments, and broad area for the big equipments [6]. With the advantages of lower capital cost, easy operation process, and high CO_2 removal percentage, membrane technology offers the best treatment for natural gas [6]. Natural gas is expected to contain less than 2 vol% or less than 2 ppm of CO_2 after the natural gas treatment in order to meet the pipeline and commercial specification [7]. This specification is made to secure the lifetime of the pipeline and to avoid an excessive budget for pipeline replacement.

Membrane technology has received significant attention from various sectors especially industries and academics in their research as it gives the most relevant impact in reducing the environmental problem and costs. Membrane is defined as a thin layer, which separates two phases and restricts transport of various chemicals in a selective manner [8]. Membrane restricts the penetration of some molecules that have bigger kinetic diameter. The commercial value of membrane is determined by the membrane's transport properties which are permeability and selectivity. Major gap of the existing technologies is limited to low CO_2 loading (<15 mol%). Ideally, we required high permeability and high selectivity of membrane, but, however, most membranes exhibit high selectivity in low permeability and vice versa which make this is as a major tradeoff of membranes, and none of these technologies are yet to treat natural gas containing high CO_2 (>80 mol%) [9].

MEMBRANE TECHNOLOGY DEVELOPMENT

Early Membrane Development

Membrane technology has been started as early as in 1850 when Graham introduced the Graham's Law of Diffusion. Then, gas separation utilization in membrane technology has been commercialized in late 1900's. Permea PRISM membrane was the first commercialized gas separation membrane produced in 1980 [2]. Summary of early development of membranes is shown in Figure 1. This innovation has led to the further membrane gas separation development. A lot of studies done by the researchers for various gas separation mostly focus on the natural gas purification.

Development of membrane for CO_2/CH_4 separation has been started since early 1990's. Numbers of membranes were fabricated using different kind of materials in the early stage of this membrane gas separation. The desirable material selected must be well suited to the separation performance by which mean separation of gases works contrarily in different materials.

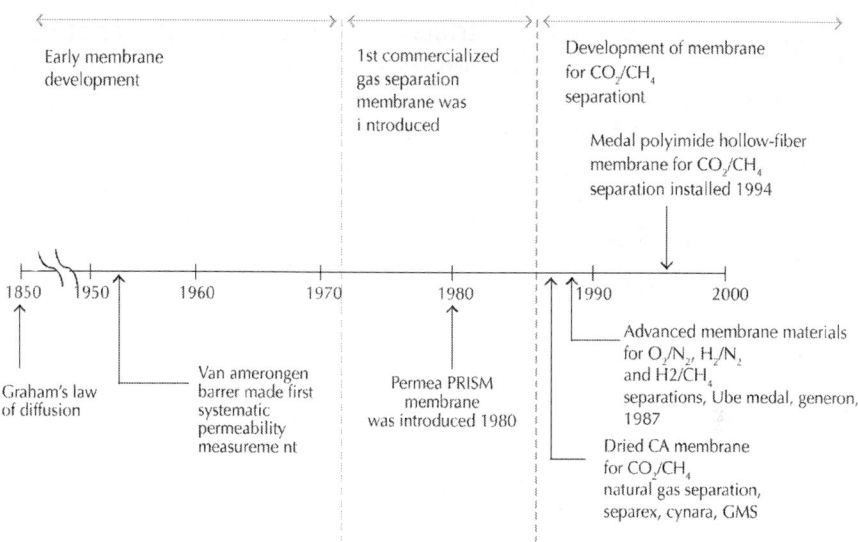

Figure 1: Membrane development timeline.

Excellent gas membranes separation should have the characteristic of high separation performance with reasonable high permeability, high robustness, chemically, thermally, and mechanically good and rational production cost [10, 11]. Two types of materials are practically used in gas separation: polymeric membrane and inorganic membrane and the comparison of both polymeric and inorganic membranes is showed in Table 2.

Table 2: Comparison between polymeric and inorganic membranes

	Polymeric membranes	Inorganic membranes
Materials	Present in either rubbery or glassy type which depends on the operating temperature [12].	Made from inorganic-based material like glass, aluminium, and metal [13].

Characteristics	Polymer is more rigid and hard in glassy state while in rubbery state it is more soft and flexible. Glassy polymeric membranes exhibit higher glass transition temperature compared to rubbery membranes, and glassy types tend to have higher CO_2/CH_4 selectivity [14].	Able to withstand with solvent and other chemicals and also susceptible to microbial attack. Comprise significantly higher permeability and selectivity, but they are also more resistant towards higher pressure and temperature, aggressive feeds, and fouling effects [15].
Disadvantages	May have plasticization problem when handling high CO_2. Presence of CO_2 may result in membrane performance reduction at certain elevated pressure. As the membranes expose to CO_2, polymer network in the membrane will swell, and segmental mobility will also increase which consequently cause a rise in permeability for all gas components [16]. The components with low permeability characteristic will experience more permeability increment; thus, the selectivity of the membrane will definitely decrease [17–19].	Inherent brittleness characteristic. Performed well under low pressure which does not suit the natural gas well which required high pressure for the exploration. High production cost which seems not practical for large industrial applications [20].
Examples	Polyethylene (PE), poly(dimethylsiloxane) (PDMS), polysulfone (PSU), polyethersulfone (PES), poly-imide (PI) [21], polycarbonate [22], polyimide [23], poly-ethers [24], polypyrrolones [25, 26], polysulfones [27], and polyethersulfones [28].	Aminoslicate membrane [29], carbon-silicalite composite membrane [30], MFI membranes [31], and microporous silica mem-branes [32].

Gas separation using polymeric membranes has taken its first commercial scale in late 1970's after the demonstration of rubbery membranes back in 1830's [33]. Literally, the permeability of gas in a specific gas mixture varies inversely with its separation factor. The tighter of molecular spacing it has, the higher the separation characteristic of the polymer, but, however, as the operating pressure increases, the permeability is decreasing due to experiencing lower diffusion coefficients [34]. Polymeric membranes that are commercially available for CO_2/CH_4 separation include polysulfone (PSU), polyetehrsulfone (PES), polyamide (PI) and many more. Generally, as the permeability of the gas increases, the permselectivity was attended to decrease in most cases of polymeric membranes [23].

Inorganic membrane like SAPO-34 could give higher separation performance compared to the polymeric membrane, but the separation performance is inversely proportional to the pressure loaded. This observation may create problem when we deal with high pressure natural gas well. The performance of both organic and inorganic membrane is summarized in Robeson's plot as in Figure 2 [35].

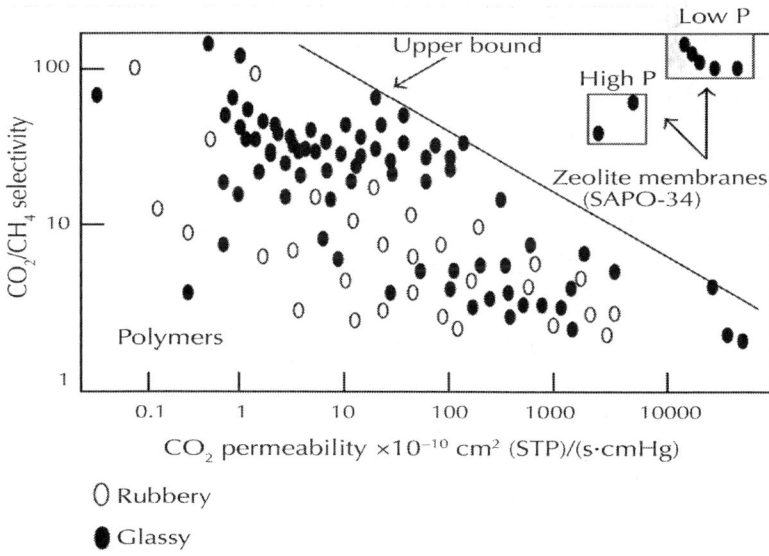

Figure 2: Zeolite (SAPO-34) membrane performance in Robeson's plot.

Conventional Mixed Matrix Membrane

A lot of researches have been done to satisfy the needs of gas separation requirement through both polymeric and inorganic membranes. The deficiencies of these membranes have driven the researchers to develop an alternative material for membrane which is more mechanically stable and economic viable, and most important is having high separation performance. The combination of organic and inorganic material which is known as mixed matrix membrane (MMM) was then proposed in idea to get a better membrane gas separation performance at reasonable price [36]. The fabrication of MMM was a promising technology as this composite material has improved its mechanical and electrical properties [37], and it combines the exceptional separation ability and pleasant stability of molecular sieves with better processability of organic membrane [38]. The MMM is characterized by dispersing the inorganic material into the continuous phase of polymeric material which can be almost any polymeric material such as polysulfone, polyimide, and polyethersulfones [39, 40].

Various membrane materials can be selected based on the process requirement. Selected materials can be "tailored-made" in order to meet the specific separation purpose in a wide range of application [39]. There were many attempts of developing polymer-inorganic membrane that started few decades back then.

Based on Table 3, this was observed that the selection of materials is important, and it depends on the system requirement. Higher intrinsic diffusion selectivity characteristic of glassy polymer makes this material better than rubbery polymer [56]. Although MMM has proven an enhancement of selectivity, it was noticed that most MMMs were endured with poor adhesion between the organic matrix and inorganic particles [55]. Even MMM fabrication does have its disadvantages, but the research of MMM with different materials is worth to work on since it has proven its ability to have high separation performance.

Table 3: Few researches of mixed matrix membranes

Year	Mixed matrix membrane (MMM)		Observations	Ref.
	Organic	Inorganic		
1973	Silicon rubber	Molecular sieves	Poor adhesion of organic and inorganic selected leads to poor separation performance. This poor interaction of both materials may result in nonselective voids present at the interface which consequently causes insufficient membrane performance [41–43].	[44]
1992	Polydimethylsilox-ane (PDMS)	Silicalite-1, 13X, KY, and zeolite-5A	Zeolite like silicalite-1, 13X, and KY have enhanced the separa-tion performance of poorly selective rubbery membrane for the carbon dioxide (CO_2) and meth-ane (CH_4) mixture.	[45]
	Propylene diene rubber (EPDM)		Zeolite-5A showed no change in gas selectivity with decrease permeabil-ity due to impermeable characteristic towards CO_2.	
2000	Cellulose acetate (CA)	Silicalite, NaX, and AgX	Silicalite did in fact reverse the selectivity of CA membrane from H_2 to CO_2 for CO_2/H_2 separa-tion.	[46]
2000	Polyvinyl acetate	4A	Formation of chemical bonds gave good adhe-sion, but there is still non-selective "leakage" from the existence of nanomet-ric region.	[47]

2003	Matrimid	Carbon molecular sieves	Selectivity of CO_2/CH_4 mixture has increased up to 45%. Zeolites loading also affects both gas permeability and gas mixture selectivity. There were also a number of records where permeability increased with selectivity decreased as the zeolites loading was increased [48, 49] and vice versa [42].	[50]
2006	Polyethersulfone (PES)	Zeolite 4A	Due to low mobility of the polymer chain in glassy polymer such as to prevent them to completely cover the zeolites surface which resulted in void interface [51, 52].	[53]
2001	Polyimide (PI)	Zeolite 13X		[54]
2008	Polycarbonate	Zeolite 4A		[55]

Recent Development of Membrane Gas Separation

Ionic Liquid-Supported Membrane (ILSM)

In recent years, many researches have been evaluated on the ionic liquid supported membrane (ILSM) for gas separation membrane since ionic liquids are known materials that could dissolve CO_2 and stable at high temperature ranges [57]. To be specific, ionic liquids are molten salt that are liquid at room temperature [58]. Furthermore, ionic liquids are of particular interest for membrane gas separation application as they are inflammable, negligible vapour pressure, and nonvolatile which make them also known as "green" solvents [58–60]. Extensive researches have been carried out to develop room temperature ionic liquid (RTIL)-based solvents for CO_2 separation with various types of ionic liquids such as pyridinium and imidazolium based. Among RTILs tested, imidazolium-based RTIL was chosen as the most feasible solvent for CO_2 separation as they are commercially viable and easily tunable by tailoring the cation and anion to meet the system requirements [60].

ILSMs have been proven that they offered an increase in permeability that outperforms many neat polymer membranes. ILSMs synthesized from poly(vinylidene fluoride) (PVDF) and 1-butyl-3-methylimidazolium tetrafluororate (BMImBF$_4$) showed high permeation performance of CO_2 and mechanically stable while operating at high pressure condition [63]. The consumption of RTILs showed an increment especially for 1-R-3-methylimidazolium (R-mim)-based RTILs as this type is preferable due to its properties of less viscous compared to other RTILs. In addition, gases like CO_2, nitrogen (N_2), and other hydrocarbons demonstrated high solubility in Rmim-based RTILs [64, 65]. Besides, the use of Rmim-based RTILs could calculate the latent permeability and selectivity of the mixture of given gases by using the molar volume of these RTILs [60]. RTIL can be functionalized and set up in according to the system requirement and application, and these researches could be good benchmark for designing the functionalized RTIL efficiently as showed in Table 4.

Table 4: Effects of ionic liquid functionalization

Functionalization	Effects
Nitrile and alkyne group	Gas solubility and separation performance have been tailored.
	Functionalized RTIL solvents displayed a decreasing in CO_2, N_2, and CH_4 solubility, but, however, the selectivity of CO_2/N_2 and CO_2/CH_4 increased when compared to the nonfunctionalized RTIL [61].
Temperature	As the temperature increases, the CO_2 solubility is decreasing while the CH_4 solubility remains unchanged.
	The ideal solubility selectivity of mix gases for CO_2/N_2, CO_2/CH_4, and CO_2/H_2 increased as the temperature decreased [62].

Polymerized Room Temperature Ionic Liquid Membrane (Poly (RTIL))

Comparatively, RTIL especially imidazolium based can be also polymerized into a solid, dense, and thin film membrane due to their modular nature [66–68]. It was a successful breakthrough when the researcher found that polymer from ionic liquid monomer had higher CO_2 absorption capacity with faster absorption and desorption rate compared to the neat RTIL [69]. Moreover, poly (RTIL) is also attributed with higher mechanical strength [66]. These characters have proven that polymerized ionic liquid (poly (RTIL)) is also a promising material for membrane gas separation. Polymerization of RTIL monomer by varying the n-alkyl length also showed a pleasant result when increase of permeability of given gases like CO_2, N_2, and methane (CH_4) was observed as the n-alkyl group was lengthened [68]. Additionally, poly (RTIL) is also up to extend when it practically absorb about twice as much CO_2 as their liquid analogue which makes it much better than molten RTIL [68]. Apparently, performance of poly (RTIL) also depends on the substituent attached to it. In a research done on the inclusion of a polar oligo(ethylene glycol) on the cation side of imidazolium-based RTIL, the separation selectivity has seemed to increase [70].

As discussed earlier, mixed matrix membrane is a known membrane that composed of a compatible organic-inorganic pair which demonstrated having good separation properties subject to no interfacial adhesion problem. The improvement of separation performance is expected in an MMM comprising poly (RTIL) (polymer matrix) and zeolite (inorganic). In a very recent work, the benefit of MMM has become an idea to the researcher in ionic liquid membrane field. Hudiono and his coworkers have introduced a three-component mixed matrix membrane by utilizing the poly (RTIL), RTIL, and zeolite [71]. Their research was also based on a positive finding by Bara and his coworkers when they found that the addition of RTIL in poly (RTIL) has increased the gas permeability. This is due to that more rapid gas diffusion occurred as the free volume of membrane increased when RTIL was added [72].

On the other hand, Hudiono has used the RTIL to increase the membrane permeability and also to act as an aid for better interaction between the polies (RTIL) and zeolite (SAPO-34). The result was promising as the permeability of given gases like CO_2, N_2, and CH_4 increased accordingly. However, the selectivity was slightly decrease as they claimed that the RTIL used which is emim[Tf_2N] was not selective towards CO_2/CH_4 separation [71]. Nonetheless, the result proved that the addition of RTIL could increase the polymer-zeolite adhesion in MMM as RTIL also acts as the wetting agent for the zeolite.

Hudiono again repeated the same experiment fabricating a three-component mixed matrix membrane but by varying the composition of RTIL and zeolite added in order to determine the optimum condition for the membrane. The CO_2 permeability seems to rise with the increasing amount of RTIL. The CO_2/CH_4 selectivity of the MMM also improved with the presence of SAPO-34 compared to neat poly (RTIL)-RTIL membrane as long as there is sufficient amount of RTIL as the wetting agent. Besides, the team also conducted an investigation of the separation performance by using the vinyl-based poly (RTIL). The addition of RTIL is not essential as they are structurally similar [73].

In contrast, a ternary MMM has been fabricated by Oral and his coworkers by using different materials. The project study on the effect of different RTIL loadings which are emim[Tf_2N] and emim[CF_3SO_3] towards MMM composed of polyimide-zeolite (SAPO-34). The addition of emim[Tf_2N] has performed as expected when the permeability of CO_2 increased while the incorporation of emim[CF_3SO_3] has increased the CO_2/CH_4 selectivity since emim[CF_3SO_3] is selective towards CO_2/CH_4 [74].

CONCLUSIONS

The escalating research in the membrane fabrication for gas separation applications signifies that membranes technology is currently growing and becoming the major focus for industrial gas separation processes. Latest research area using mixed matrix membranes combines the flexibility and low capital cost with improving selectivity, permeability, chemical, thermal, and mechanical strength. Material selection and method of preparation are the most important part in fabricating a membrane. So the next research must be very careful in determining the materials for gas separation and methods applied in the fabrication stage. Even the synthesized MMMs were only tested in a small scale, the research of MMMs is worth to be further explored since MMMs have shown better separation performance compared to polymeric and inorganic membranes.

REFERENCES

1.   Soregraph, Key World Energy Statistic, The International Energy Agency, 2010.

2.   Longterm Outlook to 2030, Natural Gas Demand and Supply, The European Union of The Natural Gas Industry, 2010.

3.   "Natural Gas and Environment—Emission from the Combustion of Natural Gas," copyright 2004–2010, http://www.naturalgas.org/environment/naturalgas.asp#emission.

4.   A. Wan and A. Rusmidah, Natural Gas, Universiti Teknologi Malaysia, 2010.

5.   D. David and D. Kishore, Recent Development in CO_2 Removal Membrane Technology, UOP, 1999.

6.   M. I. Fauzi and A. Akkil, Meeting Technical Challenge in Developing High CO_2 Gas Field Offshore, Petronas Carigali Sdn. Bhd., 2008.

7.   Fuels Providers, Natural Gas Specs Sheet, The National Petroleum Agency, 2002.

8.   Separation Process, Membrane Separation Process, Membrane Properties, 1998.

9.   Separation Process, Introduction to Membrane, Chapter 1, 1998.

10. K. Scott, Membrane Separation Technology, Scientific & Technical Information, Oxford, UK, 1990.

11. H. Strathmann, "Membrane separation processes: current relevance and future opportunities," AIChE Journal, vol. 47, no. 5, pp. 1077–1087, 2001

12. S. Morooka and K. Kusakabe, "Microporous inorganic membranes for gas separation," MRS Bulletin, vol. 24, no. 3, pp. 25–29, 1999

13. A. F. Ismail and L. I. B. David, "A review on the latest development of carbon membranes for gas separation," Journal of Membrane Science, vol. 193, no. 1, pp. 1–18, 2001

14. W. A. W. Abdul Rahman, "Formation and characterization of mixed matrix composite materials for efficient energy gas separation," Project Report, Faculty of Chemical and Natural Resources Engineering, Universiti Teknologi Malaysia, 2006

15. J. A. Ritter and A. D. Ebner, "Carbon dioxide separation technology—R&D needs for the chemical and petrochemical industries," Chemical Industry Vision 2020, 2007

16. T. Visser and M. Wessling, "When do sorption-induced relaxations in glassy polymers set in?"Macromolecules, vol. 40, no. 14, pp. 4992–5000, 2007

17. A. Bos, I. G. M. Pünt, M. Wessling, and H. Strathmann, "CO_2-induced plasticization phenomena in glassy polymers," Journal of Membrane Science, vol. 155, no. 1, pp. 67–78, 1999.

18. J. D. Wind, D. R. Paul, and W. J. Koros, "Natural gas permeation in polyimide membranes," Journal of Membrane Science, vol. 228, no. 2, pp. 227–236, 2004

19. J. D. Wind, S. M. Sirard, D. R. Paul, P. F. Green, K. P. Johnston, and W. J. Koros, "Relaxation dynamics of CO_2 diffusion, sorption, and polymer swelling for plasticized polyimide membranes,"Macromolecules, vol. 36, no. 17, pp. 6442–6448, 2003

20. A. J. Bird and D. L. Trimm, "Carbon molecular sieves used

in gas separation membranes," Carbon, vol. 21, no. 3, pp. 177–180, 1983.

21. T. H. Kim, W. J. Koros, G. R. Husk, and K. C. O'Brien, "Relationship between gas separation properties and chemical structure in a series of aromatic polyimides," Journal of Membrane Science, vol. 37, no. 1, pp. 45–62, 1988.

22. J. S. McHattie, W. J. Koros, and D. R. Paul, "Effect of isopropylidene replacement on gas transport properties of polycarbonates," Journal of Polymer Science B, vol. 29, no. 6, pp. 731–746, 1991

23. C. L. Aitken, W. J. Koros, and D. R. Paul, "Gas transport properties of biphenol polysulfones,"Macromolecules, vol. 25, no. 14, pp. 3651–3658, 1992

24. L. A. Pessan and W. J. Koros, "Isomer effects on transport properties of polyesters based on bisphenol-A," Journal of Polymer Science B, vol. 31, no. 9, pp. 1245–1252, 1993

25. D. R. B. Walker and W. J. Koros, "Transport characterization of a polypyrrolone for gas separations,"Journal of Membrane Science, vol. 55, no. 1-2, pp. 99–117, 1991

26. X. Gao, Z. Tan, and F. Lu, "Gas permeation properties of some polypyrrolones," Journal of Membrane Science, vol. 88, no. 1, pp. 37–45, 1994

27. J. S. McHattie, W. J. Koros, and D. R. Paul, "Gas transport properties of polysulphones: 2. Effect of bisphenol connector groups," Polymer, vol. 32, no. 14, pp. 2618–2625, 1991

28. Y. Liu, T. S. Chung, R. Wang, D. F. Li, and M. L. Chng, "Chemical cross-linking modification of polyimide/poly(ether sulfone) dual-layer hollow-fiber membranes for gas separation," Industrial and Engineering Chemistry Research, vol. 42, no. 6, pp. 1190–1195, 2003

29. G. Xomeritakis, C. Y. Tsai, and C. J. Brinker, "Microporous sol-gel derived aminosilicate membrane for enhanced carbon dioxide separation," Separation and Purification Technology, vol. 42, no. 3, pp. 249–257, 2005

30. L. Zhang, K. E. Gilbert, R. M. Baldwin, and J. Douglas Way,

"Preparation and testing of carbon/silicalite-1 composite membranes," Chemical Engineering Communications, vol. 191, no. 5, pp. 665–681, 2005.

31. M. P. Bernal, J. Coronas, M. Menéndez, and J. Santamaría, "On the effect of morphological features on the properties of MFI zeolite membranes," Microporous and Mesoporous Materials, vol. 60, no. 1-3, pp. 99–110, 2003

32. C. Y. Tsai, S. Y. Tam, Y. Lu, and C. J. Brinker, "Dual-layer asymmetric microporous silica membranes,"Journal of Membrane Science, vol. 169, no. 2, pp. 255–268, 2000

33. R. W. Baker, E. L. Cussler, W. Eykamp, W. J. Koros, R. L. Riley, and H. Strathmann, Membrane Separation Systems— Recent Developments and Future Directions, Noyes Data Corporation, 1991.

34. D. E. W. Vaughan, "The synthesis and manufacture of zeolites," Chemical Engineering Progress, vol. 84, no. 2, pp. 25–31, 1988

35. M. A. Carreon, Novel Membranes for Efficient CO_2 Separation, University of Lousville, 2011.

36. S. Kulprathipanja, R. W. Neuzil, and N. N. Li, "Separation of fluids by means of mixed matrix membranes in gas permeation," US Patent 4,740,219, 1988.

37. T. M. Gür, "Permselectivity of zeolite filled polysulfone gas separation membranes," Journal of Membrane Science, vol. 93, no. 3, pp. 283–289, 1994

38. L. Yi, Development of Mixed Matrix Membrane for Gas Separation Application, Tsinghua University, 2006.

39. C. M. Zimmerman, A. Singh, and W. J. Koros, "Tailoring mixed matrix composite membranes for gas separations," Journal of Membrane Science, vol. 137, no. 1-2, pp. 145–154, 1997

40. R. Mahajan, C. Zimmerman, and W. Koros, Fundamental, Practical Aspects of Mixed Matrix Gas Separation Membranes, ACS Symposium Series, 1999.

41. V. Bhardwaj, A. MacIntosh, I. D. Sharpe, S. A. Gordeyev,

and S. J. Shilton, "Polysulfone hollow fiber gas separation membranes filled with submicron particles," Annals of the New York Academy of Sciences, vol. 984, pp. 318–328, 2003

42. R. Mahajan, R. Burns, M. Schaeffer, and W. J. Koros, "Challenges in forming successful mixed matrix membranes with rigid polymeric materials," Journal of Applied Polymer Science, vol. 86, no. 4, pp. 881–890, 2002

43. M. G. Süer, N. Baç, and L. Yilmaz, "Gas permeation characteristics of polymer-zeolite mixed matrix membranes," Journal of Membrane Science, vol. 91, no. 1-2, pp. 77–86, 1994

44. D. R. Paul and D. R. Kemp, "The diffusion time lag in polymer membrane containing adsorptive fillers,"Journal of Polymer Science C, no. 41, pp. 79–93, 1973

45. J. M. Duval, B. Folkers, M. H.V. Mulder, G. Desgrandchampsb, and C. A. Smolders, "Adsorbent filled membranes for gas separation. Part 1. Improvement of the gas separation properties of polymeric membranes by incorporation of microporous adsorbents," Journal of Membrane Science, vol. 80, no. 1, pp. 189–198, 1992

46. S. Kulprathipanja, "Review of recent progress in mixed matrix membranes," Membrane Technology, vol. 105, pp. 6–8, 2000

47. R. Mahajan and W. J. Koros, "Factors controlling successful formation of mixed-matrix gas separation materials," Industrial and Engineering Chemistry Research, vol. 39, no. 8, pp. 2692–2696, 2000

48. J. M. Duval, Adsorbent filled polymeric membranes [Ph.D. thesis], The University of Twente, 1995.

49. Z. Huang, J. F. Su, X. Q. Su, Y. H. Guo, L. J. Teng, and C. M. Yang, "Preparation and permeation characterization of -zeolite-incorporated composite membranes," Journal of Applied Polymer Science, vol. 112, no. 1, pp. 9–18, 2009

50. D. Q. Vu, W. J. Koros, and S. J. Miller, "Mixed matrix membranes using carbon molecular sieves: I. Preparation and experimental results," Journal of Membrane Science, vol.

211, no. 2, pp. 311–334, 2003.

51.   M. D. Jia, K. V. Peinemann, and R. D. Behling, "Preparation and characterization of thin-film zeolite-PDMS composite membranes," Journal of Membrane Science, vol. 73, no. 2-3, pp. 119–128, 1992

52.   T. W. Pechar, S. Kim, B. Vaughan et al., "Preparation and characterization of a poly(imide siloxane) and zeolite L mixed matrix membrane," Journal of Membrane Science, vol. 277, no. 1-2, pp. 210–218, 2006.

53.   Z. Huang, Y. Li, R. Wen, M. M. Teoh, and S. Kulprathipanja, "Enhanced gas separation properties by using nanostructured PES-zeolite 4A mixed matrix membranes," Journal of Applied Polymer Science, vol. 101, no. 6, pp. 3800–3805, 2006

54.   H. H. Yong, H. C. Park, Y. S. Kang, J. Won, and W. N. Kim, "Zeolite-filled polyimide membrane containing 2,4,6-triaminopyrimidine," Journal of Membrane Science, vol. 188, no. 2, pp. 151–163, 2001.

55.   D. Sen, Polycarbonate based zeolite 4A filled mixed matrix membranes: preparation, characterization and gas separation performances [Ph.D. thesis], Middle East Technical University, 2008.

56.   D. R. Paul and D. R. Kemp, "Diffusion time lag in polymer membranes containing adsorptive fillers,"Journal of Polymer Science C, no. 41, pp. 79–93, 1973

57.   J. D. Figueroa, T. Fout, S. Plasynski, H. McIlvried, and R. D. Srivastava, "Advances in CO_2 capture technology-The U.S. Department of Energy's Carbon Sequestration Program," International Journal of Greenhouse Gas Control, vol. 2, no. 1, pp. 9–20, 2008

58.   M. Smiglak, W. M. Reichert, J. D. Holbrey et al., "Combustible ionic liquids by design: is laboratory safety another ionic liquid myth?" Chemical Communications, no. 24, pp. 2554–2556, 2006

59.   M. J. Earle, J. M. S. S. Esperança, M. A. Gilea et al., "The distillation and volatility of ionic liquids,"Nature, vol. 439,

no. 7078, pp. 831–834, 2006

60.   D. Camper, J. Bara, C. Koval, and R. Noble, "Bulk-fluid solubility and membrane feasibility of Rmim-based room-temperature ionic liquids," Industrial and Engineering Chemistry Research, vol. 45, no. 18, pp. 6279–6283, 2006

61.   T. K. Carlisle, J. E. Bara, C. J. Gabriel, R. D. Noble, and D. L. Gin, "Interpretation of CO_2 solubility and selectivity in nitrile-functionalized room-temperature ionic liquids using a group contribution approach," Industrial and Engineering Chemistry Research, vol. 47, no. 18, pp. 7005–7012, 2008

62.   A. Finotello, J. E. Bara, D. Camper, and R. D. Noble, "Room-temperature ionic liquids: temperature dependence of gas solubility selectivity," Industrial and Engineering Chemistry Research, vol. 47, no. 10, pp. 3453–3459, 2008

63.   Y. I. Park, B. S. Kim, Y. H. Byun, S. H. Lee, E. W. Lee, and J. M. Lee, "Preparation of supported ionic liquid membranes (SILMs) for the removal of acidic gases from crude natural gas," Desalination, vol. 236, no. 1-3, pp. 342–348, 2009

64.   D. Camper, C. Becker, C. Koval, and R. Noble, "Low pressure hydrocarbon solubility in room temperature ionic liquids containing imidazolium rings interpreted using regular solution theory,"Industrial and Engineering Chemistry Research, vol. 44, no. 6, pp. 1928–1933, 2005

65.   P. Scovazzo, J. Kieft, D. A. Finan, C. Koval, D. DuBois, and R. Noble, "Gas separations using non-hexafluorophosphate [PF6]- anion supported ionic liquid membranes," Journal of Membrane Science, vol. 238, no. 1-2, pp. 57–63, 2004

66.   H. Ohno, M. Yoshizawa, and W. Ogihara, "Development of new class of ion conductive polymers based on ionic liquids," Electrochimica Acta, vol. 50, no. 2-3, pp. 255–261, 2004

67.   X. Hu, J. Tang, A. Blasig, Y. Shen, and M. Radosz, "CO_2 permeability, diffusivity and solubility in polyethylene glycol-grafted polyionic membranes and their CO_2 selectivity relative to methane and nitrogen," Journal of Membrane Science, vol. 281, no. 1-2, pp. 130–138, 2006

68. J. E. Bara, S. Lessmann, C. J. Gabriel, E. S. Hatakeyama, R. D. Noble, and D. L. Gin, "Synthesis and performance of polymerizable room-temperature ionic liquids as gas separation membranes," Industrial and Engineering Chemistry Research, vol. 46, no. 16, pp. 5397–5404, 2007.

69. J. Tang, W. Sun, H. Tang, M. Radosz, and Y. Shen, "Enhanced CO_2 absorption of poly(ionic liquid)s,"Macromolecules, vol. 38, no. 6, pp. 2037–2039, 2005.

70. J. E. Bara, C. J. Gabriel, S. Lessmann et al., "Enhanced CO_2 separation selectivity in oligo(ethylene glycol) functionalized room-temperature ionic liquids," Industrial and Engineering Chemistry Research, vol. 46, no. 16, pp. 5380–5386, 2007.

71. Y. C. Hudiono, T. K. Carlisle, J. E. Bara, Y. Zhang, D. L. Gin, and R. D. Noble, "A three-component mixed-matrix membrane with enhanced CO_2 separation properties based on zeolites and ionic liquid materials," Journal of Membrane Science, vol. 350, no. 1-2, pp. 117–123, 2010.

72. J. E. Bara, D. L. Gin, and R. D. Noble, "Effect of anion on gas separation performance of polymer-room-temperature ionic liquid composite membranes," Industrial and Engineering Chemistry Research, vol. 47, no. 24, pp. 9919–9924, 2008.

73. Y. C. Hudiono, T. K. Carlisle, A. L. LaFrate, D. L. Gin, and R. D. Noble, "Novel mixed matrix membranes based on polymerizable room-temperature ionic liquids and SAPO-34 particles to improve CO_2 separation," Journal of Membrane Science, vol. 370, no. 1-2, pp. 141–148, 2011.

74. C. A. Oral, R. D. Noble, and S. B. Tantekin-Ersolmaz, "Ternary mixed-matrix membranes containing room temperature ionic liquids," in Proceedings of the North American Membrane Society Conference (NAMS '11), 2011.

Chapter 6

Parametric Investigation of Well Testing Analysis in Low Permeability Gas Condensate Reservoirs

Hossein Mohammadi[a], Mohammad Hossein Sedaghat[b], and Abbas Khaksar Manshad[c]

[a]Faculty of Engineering, Department of Chemical Engineering, Bushehr Branch, Islamic Azad University, Ali Shahr Bushehr, Iran
[b]Faculty of Chemical Engineering, Dashtestan Branch, Islamic Azad University, Dashtestan, Iran
[c]Faculty of Engineering, Department of Chemical Engineering, Persian Gulf University, Bushehr, Iran

ABSTRACT

Most of gas condensate reservoirs (GCRs) are detected in deep and tight formations. In these reservoirs, formation of near wellbore gas condensate is a significant factor influencing the production even in low fluid richness situation. The formation of condensate jots while the pressure dropping below the dew point and low well deliverability in low permeability GCRs complicate analyzing the well test and characterizing the reservoir. In this work, the pseudo-pressure functions were applied to linearize a two-phase flow including gas and gas condensate through the porous media. In analyzing the well test, various techniques such as a single-phase function and a two-phase pseudo-pressure function were used. In addition, a two-zone steady-state and a three-zone method were examined to calculate a two-phase pseudo-pressure function. Also, the role of different parameters e.g. fluid richness, relative permeability curves, mechanical skin factor, initial pressure difference with the dew point and flow rate were precisely determined. The effects of various causes on the condensate formation and pseudo-pressure function were studied and the accuracy of permeability and skin factor estimations of mentioned various methods in different near wellbore regions were determined.

INTRODUCTION

Synthetic well test data generated from a compositional simulator were used to analyze the build-up tests in low permeability GCR models. For wells producing at a constant rate during the transient flow period in which the change rate of the reservoir pressure with respect to time is not zero or constant, the use of the pseudo-pressure approach was reviewed. Some present analysis methods such as, the single-phase dry gas pseudo-pressure, the two-zone steady-state, and the three-zone were compared with each other. The applicability of the dry gas single-phase pseudo-pressure function as a fast estimation technique was examined and it did not result in linear diffusivity equations (as predicted by Mazloom and Rashidi, 2006). Consequently, the two-phase pseudo-pressure function for linearizing the flow of gas condensate in porous media was used as the theoretical basis to analyze the well test data presented in this study.

In most GCRs, the well bottom-hole as well as the near-wellbore pressures drop below the dew point pressure. These reservoirs tend to exhibit a complex behavior due to existence of a two-phase system, i.e. gas and the liquid condensate. As a result, the analysis of pressure transient in gas condensate wells is strongly influenced by the presence of the liquid phase in the proximity of the wellbore (Bemani et al., 2003).

It has been recently found that as the pressure of a reservoir around a well drops below the dew point pressure, retrograde condensation occurs and four regions are created with different liquid saturations (Gringarten et al., 2006). These are: (1) an outer region away from the wellbeing above the dew point pressure, contains gas with the initial liquid saturation; (2) an intermediate region with a rapid increase in liquid saturation and a corresponding decrease in gas relative permeability where the liquid condensate is immobile and only the gas is flowing; (3) a closer region to the well forms where the liquid saturation reaches a critical value, and the effluent travels as a two-phase fluid with constant composition (the deposited condensate is equal to that flowed toward the well when the pressure decreases; and the speed of condensate deposition is the same as condensate well flux rate and according to O'Dell and Miller (1967) there is no accumulation, so the flow is at a steady-state); (4) velocity stripping zone, in the immediate vicinity of the well indicates an increase in gas relative permeability compared with the gas relative permeability of Region 3, and more entrained condensates are delivered by the gas phase. Therefore, it is characterized by a decrease in liquid saturation. The Region 4 is developed at low interfacial tensions (due to more miscibility of liquid and gas phases) or at high rates. These regions are illustrated in Fig. 1.

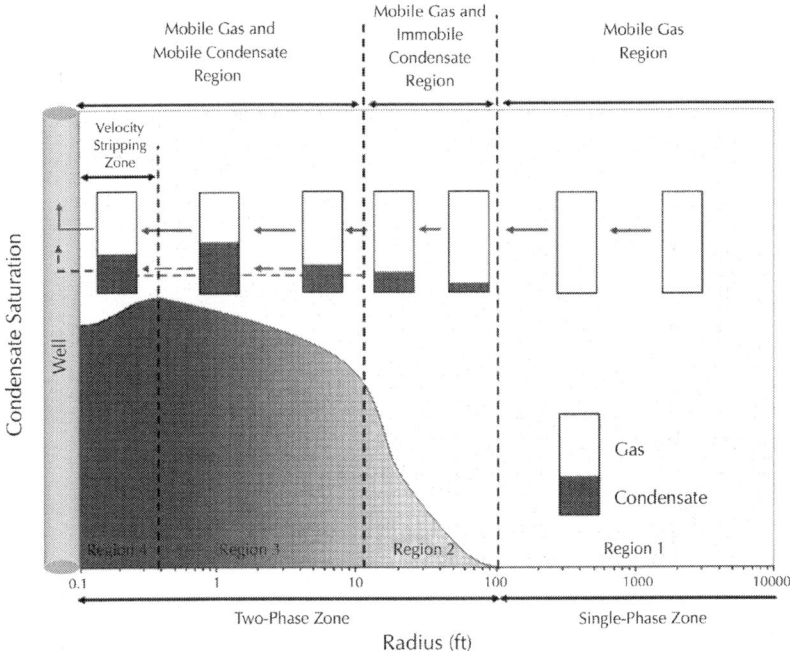

Figure 1: Condensate saturation profile with condensate dropout and velocity stripping.

O'Dell and Miller (1967) presented a simple method to evaluate the well deliverability for a GCR based on steady-state flow concepts. This method is valid when the produced well stream is the original reservoir gas and when the blockage radius is relatively small. Fussell (1973) used equation of state (EOS) compositional simulation to study the deliverability of radial gas condensate wells produced by pressure depletion below the dew point. He showed that the O'Dell and Miller's (1967) equation, with a small correction, considering dissolved gas in the flowing oil phase, dramatically overpredicts the deliverability loss resulted from condensate blockage comparing with the deliverability loss of simulation results.

Jones et al. (1989) presented a relation between the pressure and saturation to compute the pseudo-pressure based on the O'Dell and Miller's (1967) steady-state flow model. This theory states that the ratio k_{rc}/k_{rg} for steady-state flow model in two-phase regions is shown through the following equation by Jones et al. (1989):

$$\frac{k_{rc}}{k_{rg}} = \frac{\rho_g L \mu_g}{\rho_c V \mu_c}$$

(1)

where k_{rc}: condensate relative permeability, unitless; k_{rg}: gas relative permeability, unitless; ρ_g: gas density, lb/ft³; ρ_c: condensate density, lb/ft³; μ_g: gas viscosity, cp; μ_c: condensate viscosity, cp.

L and V are the molar fractions of liquid and gas phases respectively calculated from flash calculation in constant composition expansion (CCE) experiment.

The steady-state two-zone pseudo-pressure is calculated through the following equation (Jones et al., 1989):

$$m^{2Z}(p) = 2 \left[\int_{p_{wf}}^{p_{dew}} \left(\frac{k_{rg}}{\mu_g Z_g} + \frac{k_{rc}}{\mu_c Z_c} \right) p dp + k_{rg}(S_{wc}) \int_{p_{dew}}^{p_i} \left(\frac{1}{\mu_g Z_g} \right) p dp \right]$$

(2)

where $m^{2Z}(p)$: two-zone pseudo-pressure, psi²/cp; p_{wf}: well flowing pressure, psi; p_{dew}: dew point pressure, psi; Z_g: gas compressibility factor, unitless; Z_c: condensate compressibility factor, unitless; $k_{rg}(S_{wc})$: gas relative permeability at connate water saturation, unitless; p_i: initial reservoir pressure, psi.

The Steady-state flow model assumes a radial composite GCR including only Regions 1 and 3. The first integral of Eq. (2) represents two-phase pseudo-pressure in Region 3, for the pressure range $[p_{wf}, p_{dew}]$, and the second integral carries Al-Hussainy and Ramey's (1966) equation for single-phase pseudo-pressure in Region 1, for the pressure range $[p_{dew}, p_i]$, while changes occur to this equation because of the effect of S_{wc}.

A fundamental understanding of gas condensate well deliverability was made by Fevang and Whitson (1995). They developed a simple method using a two-phase pseudo-pressure and a three region model to calculate gas condensate well production performance. The two-phase three-zone pseudo-pressure is defined as (Fevang and Whitson, 1995):

$$m^{3Z}(p) = 2\left[\int_{p_{wf}}^{p^*} \left(\frac{k_{rg}}{\mu_g Z_g} + \frac{k_{rc}}{\mu_c Z_c}\right)pdp + \int_{p^*}^{p_{dew}} \left(\frac{k_{rg}}{\mu_g Z_g}\right)pdp \right.$$

$$\left. + k_{rg}(S_{wc}) \int_{p_{dew}}^{p_i} \left(\frac{1}{\mu_g Z_g}\right)pdp \right]$$

(3)

where $m^{3Z}(p)$ and p^* are the three-zone pseudo-pressure and the external pressure of the Region 3, respectively.

The first and the third integrals of Eq. (3) stand for Region 3, in the pressure range $[p_{wf}, p^*]$, and for Region 1, respectively. In Region 3 the ratio k_{rc}/k_{rg} is calculated in a different way. In addition, the second integral assumes the effect of Region 2, for the pressure range $[p^*, p_{dew}]$.

Fevang and Whitson (1995) showed that gas condensate well rate can be easily calculated by using the instantaneous producing gas-condensate ratio (generated from simulators), fluid PVT, and gas-oil relative permeabilities. The ratio k_{rc}/k_{rg} as a function of pressure in Region 3 is calculated from the following equation (Fevang and Whitson, 1995):

$$\frac{k_{rc}}{k_{rg}} = \left(\frac{1 - r_s R_p}{R_p - R_s}\right) \frac{\mu_c B_c}{\mu_g B_g}$$

(4)

where r_s: condensate capacity of entrance gas to Region 3, stb/Mscf; R_p: instantaneous gas/condensate ratio, Mscf/stb; R_s: solution gas/condensate ratio, Mscf/stb; B_c: condensate formation volume factor, bbl/stb; B_g: gas formation volume factor, ft³/scf.

The liquid condensate of Region 2 is immobile, $k_{rc} = 0$. The gas phase relative permeability, k_{rg}, is calculated based on a function of condensate saturation. The condensate saturation in Region 2, is estimated as function of pressure from the constant volume depletion (CVD) experiment corrected for the connate water saturation (Fevang and Whitson, 1995):

$$S_C(p) = [V_{rcCVD}(p)(1 - S_{wc})]$$

(5)

where $S_c(p)$ stands for the condensate saturation and $V_{rcCVD}(p)$ stands for the relative condensate volume obtained from CVD.

In 1999, Xu and Lee (1999a) investigated the gas condensate problem and tried to improve previous solutions suggested the two-zone, radial composite case. Finally the work by Xu and Lee (1999b) presented a three-zone radial composite reservoir model. The first zone assumed that only dry gas exists in this region. The second zone assumed immobile condensate saturation while there was a mobile gas phase in it. The third zone near the well assumed the steady-state two-phase flow. The behavior of these zones was the same as described in the work by Fevang and Whitson (1995).

In the flow of gas condensate fluids through porous media at high velocities, two competing phenomena appeared which might make the effective gas permeability be rate dependent which were (Jamiolahmady et al., 2003): (1) An increase in relative permeability with Darcy velocity. This effect was sometimes termed 'velocity stripping' or 'positive coupling' and (2) Inertial (non-Darcy) flow effects, which reduced the effective gas permeability at high velocity. These two high velocity phenomena acted in opposite directions and developed a forth region just near the wellbore.

The ratio k_{rc}/k_{rg} in Region 4 can be calculated by using the procedure described by Bonyadi et al. (2012), in which the effect of velocity striping is modeled through including capillary number effect in the relative permeability curves. They introduced a Darcy velocity dependent multiplication factor to model the inertial flow effects.

In this study, 35 cases were considered in compositional simulation runs and the effect of rock and fluid parameters (e.g. fluid richness, relative permeability curves and mechanical skin), initial pressure difference and production rate, on the well test behavior of low permeability GCRs were accurately investigated and the applicability of single-phase and two-phase pseudo-pressure methods were discussed. In the case of Region 4 presence, the calculation of two-phase pseudo-pressure became velocity dependent and it was not taken into account in this work. And the grid block sizes and rates were carefully selected to keep capillary number and velocity low enough to avoid formation of this region around the wellbore.

DESCRIPTION OF MODEL

To perform PVT experiments, the PVTi module of ECLIPSE was used. PVTi is a compositional pressure volume temperature (PVT) equation of state-based program used for characterizing a set of fluid samples to use in ECLIPSE simulators. The data for the PVTi program is the composition of synthetic reservoir fluid in Region 1 and temperature as shown in Table 1.

Table 1: Synthetic mixture compositions

Mixture	Fluid type	T (°F)	P_{dew} (Psi)	L^{CCE}_{max} (%)	Composition (%)		
					Methane	n-bu-tane	Decane
Mix1	Lean	140	4116	5.7	95.6	1.8	2.6
Mix2	Intermedi-ate	260	4512	13.7	89.6	3	7.4
Mix3	Rich	220	5269	23. 9	89	1.55	9.45
Mix4	Near criti-cal	220	5380	33.7	87	1.5	11.5

Fig. 2 represents the CCE condensate dropouts as predicted by the three parameter Peng–Robinson equation of state. The figure clearly suggests a low, intermediate and high liquid dropout of mixtures 1, 2 and 3 (2.6%, 7.4% and 9.45% C_{10}) typical of a lean, intermediate and a rich sample, while mixture 4 (11.5% C_{10}) exhibits a very large liquid dropout just a few psi below dew point pressure, characterizing the fluid as the near critical.

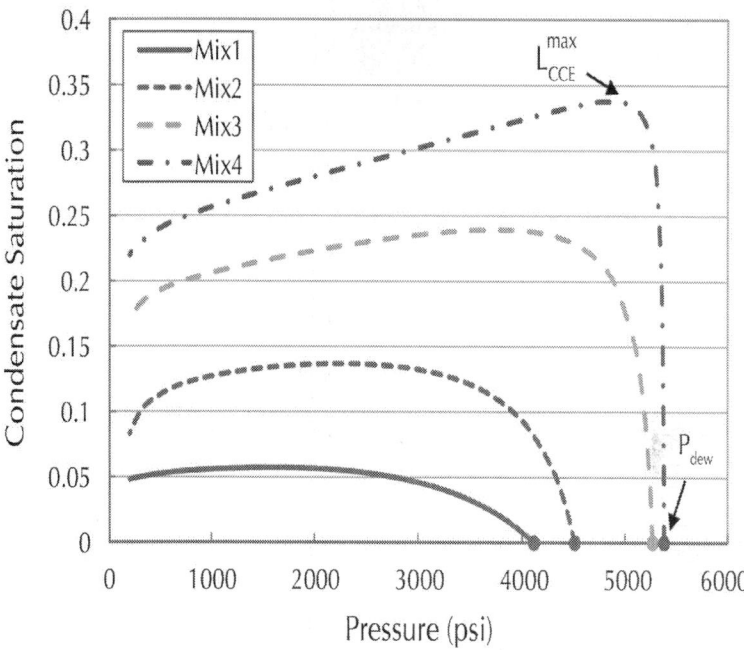

Figure 2: CCE experiment performed on four studied mixtures.

To prepare enough pressure transient data for properly detection of all regions in the well testing analysis, the reservoir fluid was considered at an initial condition above dew point pressure as the gaseous phase. The validity of well test pressure responses obtained from simulation, were analyzed. The reservoir permeability and skin were calculated and compared with the input data in the simulation model to verify the accuracy of the simulation model.

The fine grid cells near the wellbore were used to properly model the behavior of the reservoir due to the buildup of condensate near the wellbore. We maintained the use of coarser grids for the remainder of the reservoir. The reservoir is divided into 80 grid blocks in the radial direction using the log-spacing cell sizes (Table 2). The permeability of a particular class of low permeability GCR seldom exceeds 10 mD; hence the typical values of 3, 5 and 7 mD were used in the sensitivity study. The main parameters used for simulation of all cases are listed in Table 3.

Table 2: Cell sizes in radial direction

Number of cells	80
Wellbore cell radius (ft)	0.25
Cell sizes (ft)	0.29, 0.37, 0.39, 0.45, 0.53, 0.62, 0.71, 0.83, 0.96, 1.12, 1.31 1.51, 1.76, 2.04, 2.37, 2.75, 3.21, 3.72, 4.32, 5.02, 5.83, 6.78, 7.88, 9.15, 10.63, 12.35, 14.35, 16.67, 19.37, 22.51, 26.15, 30.38, 35.29, 41, 47.64, 55.35, 64.31, 74.72, 86.81, 100.86, 117.18, 136.14, 158.18, 183.78, 213.52, 248.1, 288.22, 334.86, 389.05, 452.01, 525.16, 610.15, 708.89, 823.62, 956.91, 1111.76, 1291.69, 1500.73, 1743.6, 5*1800, 16*2000

Table 3: Main simulation parameters

Parameter	Base case	Sensitivity study
Average porosity (%)	20	10–30
Absolute permeability (mD)	5	3–7
Formation thickness (ft)	200	–
$p_i - p_{dew}$ (psi)	200	50–200
Wet gas rate (MMscf/day)	10	8–15
Mechanical skin	0	–1–3
Draw-down time	60 days	5 h–60 days
Mixture	Mix3	Mix1–Mix4

Corey's relative permeability correlation has been found useful for describing low capillary number behavior of steady-state data measured for gas condensate cores. The imbibition relative permeability curves used in the simulation are based on Corey's expression in Eqs. (6) and (7) (Liu et al., 2001).

$$k_{rg} = k_{rg}^{max}(S_{wc})\left(S_g^*\right)^{n_g}$$

(6)

$$k_{rc} = k_{rc}^{max}(S_{wc})\left(S_c^*\right)^{n_g}$$

(7)

where S_g^* and S_c^* are gas and condensate normalized saturations, respectively and are defined as Eqs. (8) and (9).

$$S_g^* = \frac{S_g - S_{gc}}{1 - S_{gc} - S_{wc}}$$

(8)

$$S_c^* = \frac{S_c - S_{cc}}{1 - S_{cc} - S_{wc}}$$

(9)

Corey's correlation parameters presented in Table 4 have been studied to cover a wide range of relative permeability cases possibly encountered in GCRs. Sixteen cases of 35 simulation runs studied in this article are relative permeability curve sensitivity analysis cases, while other cases are considered to study the effect of fluids, mechanical skin, pressure difference between initial reservoir pressure and dew point pressure and wet gas rate.

Table 4: Corey's correlation parameters

Parameter	Base case	Sensitivity study
Gas end-point relative permeability	1	0.5–1
Condensate end-point relative permeability	1	0.4–1
Gas relative permeability exponent	3	1–3
Condensate relative permeability exponent	3	1–3
Connate water saturation (%)	0	0–30
Critical condensate saturation (%)	20	0–30
Critical gas saturation (%)	0	–

The skin is modeled by a radial zone of lower permeability around the well (Jokhio, 2002). The total skin in a gas condensate well is given by the sum of four individual components: (1) Flow restrictions from mechanical damage, S_{damage}, (2) Flow restriction from partial completion, $S_{partial}$, (3) Flow restriction from non-Darcy flow effects, $S_{non-Darcy}$, (4) Flow restriction from permeability reduction due to liquid dropout, $S_{liquid-dropout}$. The interaction of partial completion and non-Darcy is not considered in this work. Mathematically the total apparent skin factor for condensate well test interpretation is given in Eq. (10).

$$S_t = S_{damage} + S_{partial} + S_{non-Darcy} + S_{liquid-dropout}$$
(10)

Sixty days' draw-down following by 100 days' buildup was performed to produce synthetic well test data. Obtained well test data was fed into a developed program to calculate single-phase pseudo-pressure and two-phase pseudo-pressure by steady-state and three-zone methods.

RESULTS AND DISCUSSIONS

Description of the Cases Studied

Before studying the simulation and well testing results, it is necessary to present a brief description of the all cases under consideration.

Reservoir Fluid Mixture

Four simple methane–butane–decane mixtures (C_1–C_4–C_{10}) were considered for demonstration purposes. The mixtures have different composition and behave as retrograde gas condensate systems at the associated reservoir temperature and pressure range. Their compositions have been chosen such that Mix1 corresponds to a lean gas condensate system and Mix4 to a near critical system, Mix2 and Mix3 are intermediate and rich, respectively. The liquid dropout during a CCE experiment was presented in Fig. 2. A richer system has a higher liquid dropout in a CCE and a higher condensate-gas ratio at dew point.

Relative Permeability Curves

Sixteen sets of relative permeability curves with parameters discussed in Table 4 are considered. As illustrated in Fig. 6, these sets are selected to show the effect of S_{wc}, S_{cc}, n_g and n_c, K_{rg}^{max} and K_{rc}^{max} on the well testing behavior of gas condensate systems.

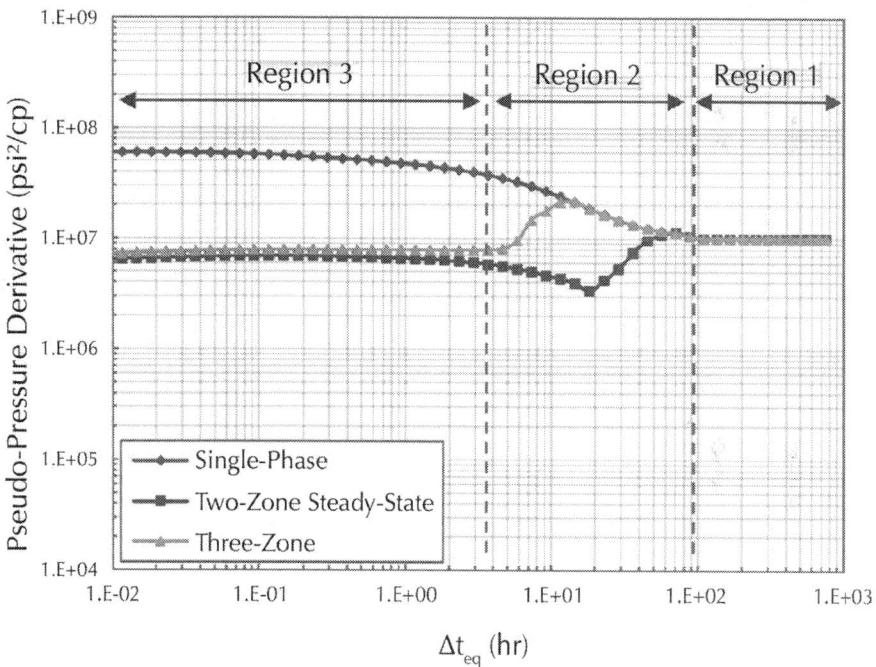

Figure 3: Comparison of typical single-phase and two-phase pseudo-pressure derivatives plots.

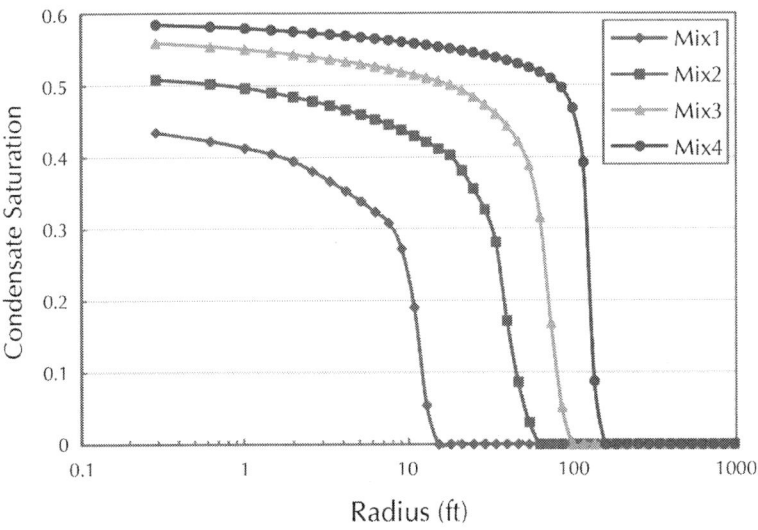

Figure 4: Condensate dropout profile in the reservoir at shut-in time, richness effect.

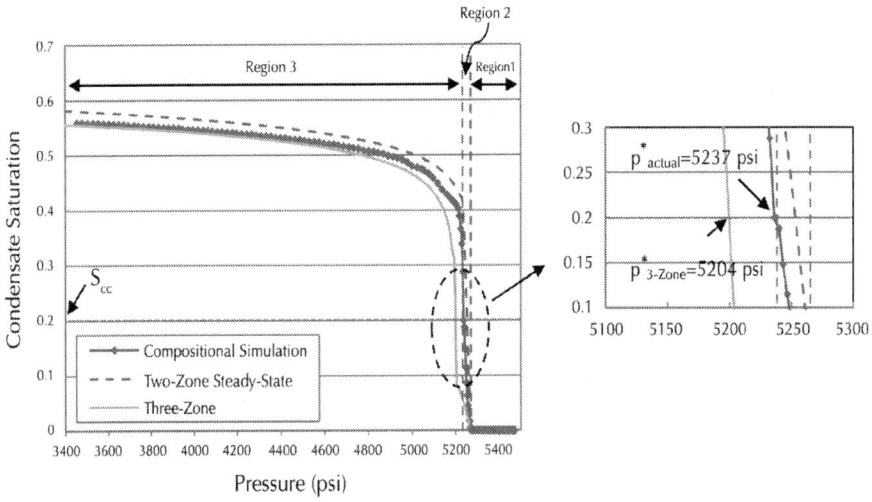

Figure 5: Comparison of condensate saturation from the simulation and as predicted by three-zone and steady-state methods at shut-in time for Mix3.

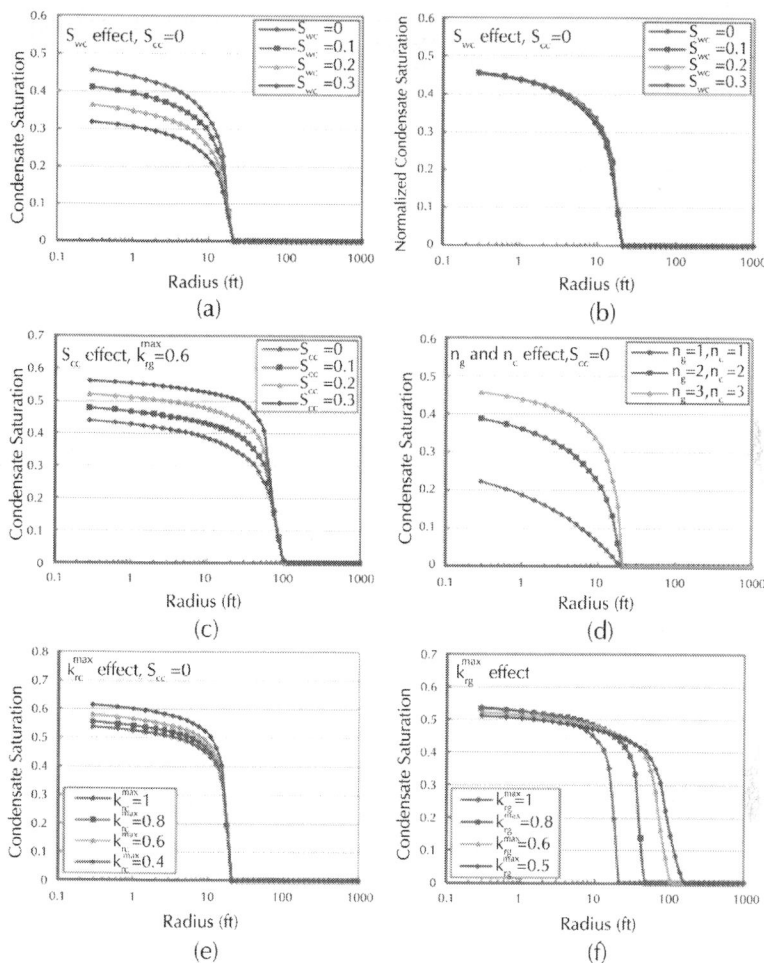

Figure 6: Condensate dropout profile in the reservoir: (a) S_{wc} effect for S_{cc} = 0, (b) its normalized saturation plot, (c) S_{cc} effect for K_{rg}^{max} = 0.6, (d) n_g and n_c effect for S_{cc} = 0, (e) K_{rc}^{max} effect for S_{cc} = 0 and (f) K_{rg}^{max} effect.

Mechanical Skin

Four different values for skin are considered: −1, 0, 1.5 and 3, the corresponding impacts of the skin zone are shown in Fig. 8.

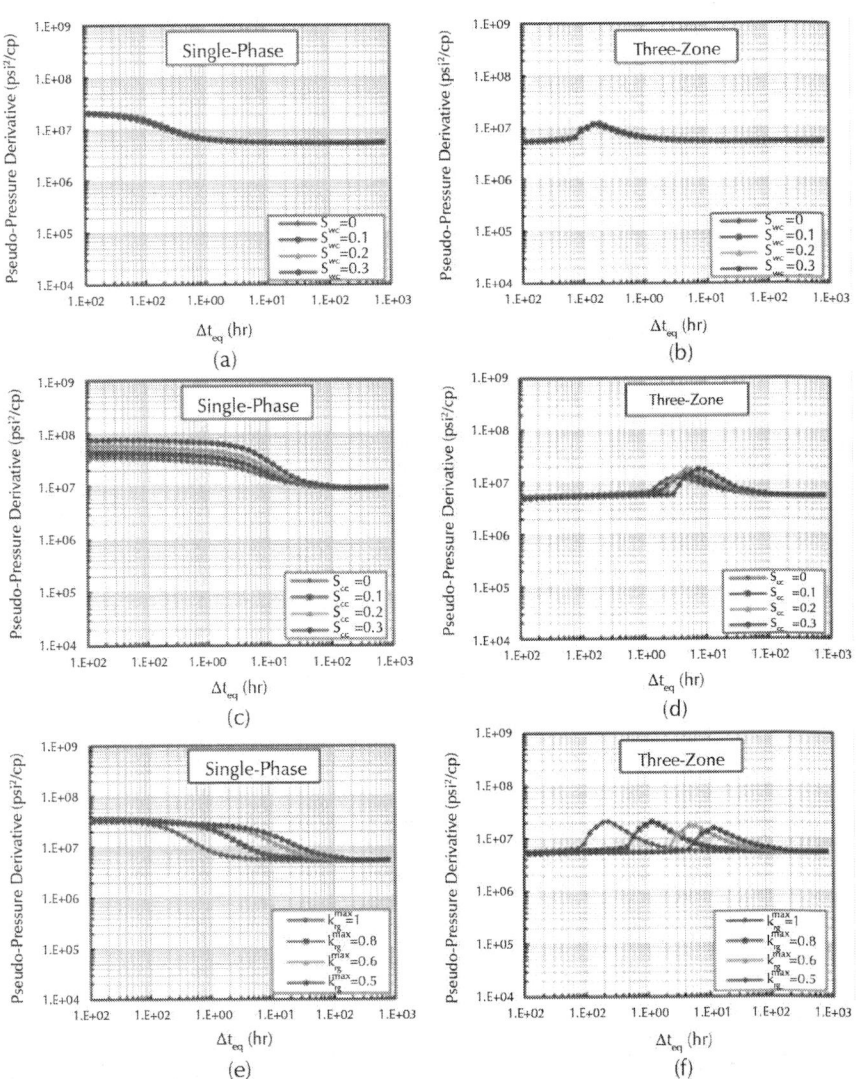

Figure 7: Buildup analysis, single-phase and two-phase three-zone pseudo-pressure derivative, (a) and (b)S_{wc} effect, (c) and (d) S_{cc} effect, (e) and (f) K_{rg}^{max} effect.

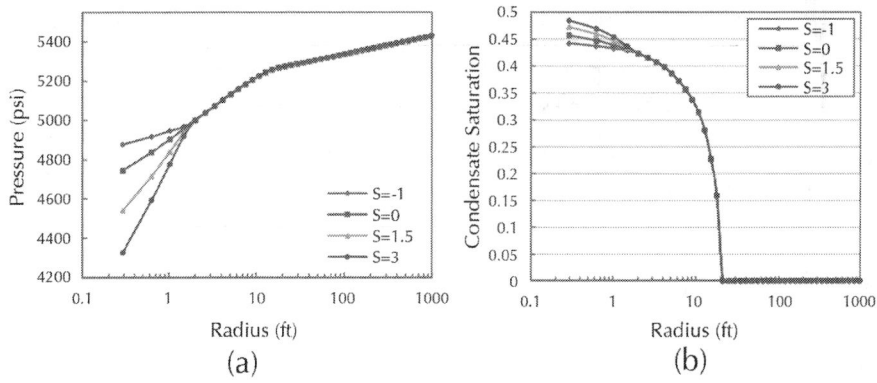

Figure 8: (a) Pressure and (b) condensate dropout profile in the reservoir at shut-in time, mechanical skin effect.

Initial Pressure Difference (p_i – pdew)

Three different initial pressure differences are tested: 200, 150, and 100 psi. The effect of $p_i - p_{dew}$ on condensate dropout is illustrated in Fig. 11.

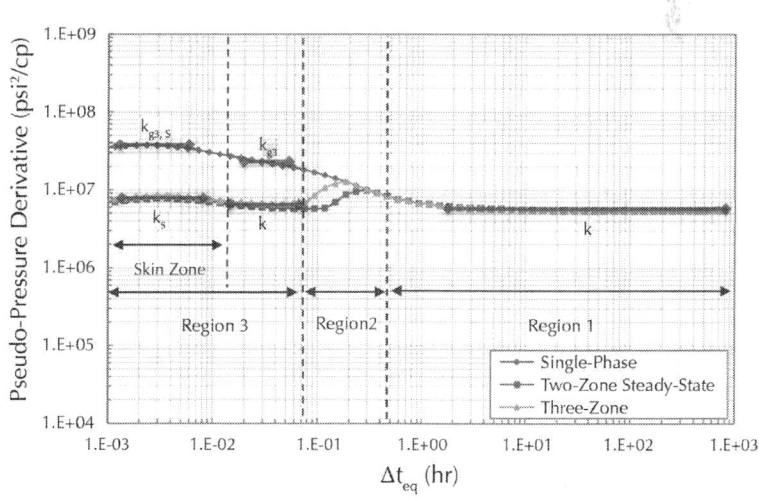

Figure 9: Comparison of three-zone and steady-state two-phase pseudo-pressure derivative with single-phase pseudo-pressure derivative, $S = 1.5$.

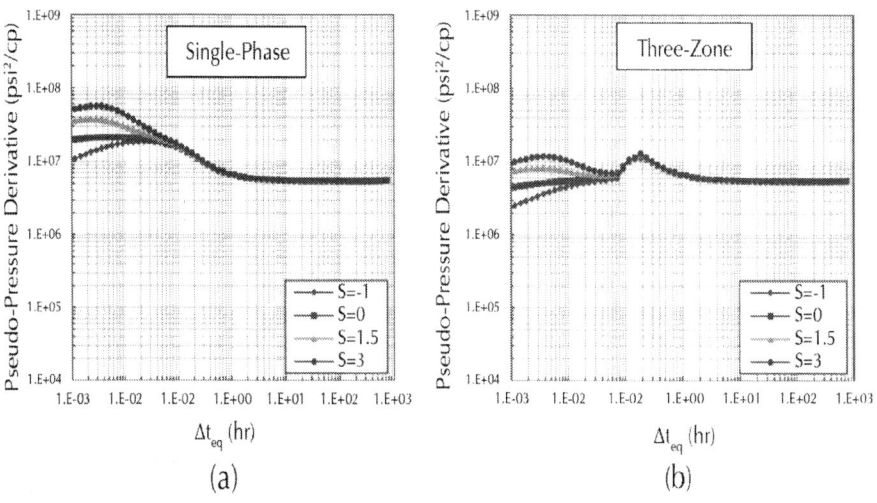

Figure 10: Buildup analysis, (a) single-phase and (b) two-phase three-zone pseudo-pressure derivative, mechanical skin effect.

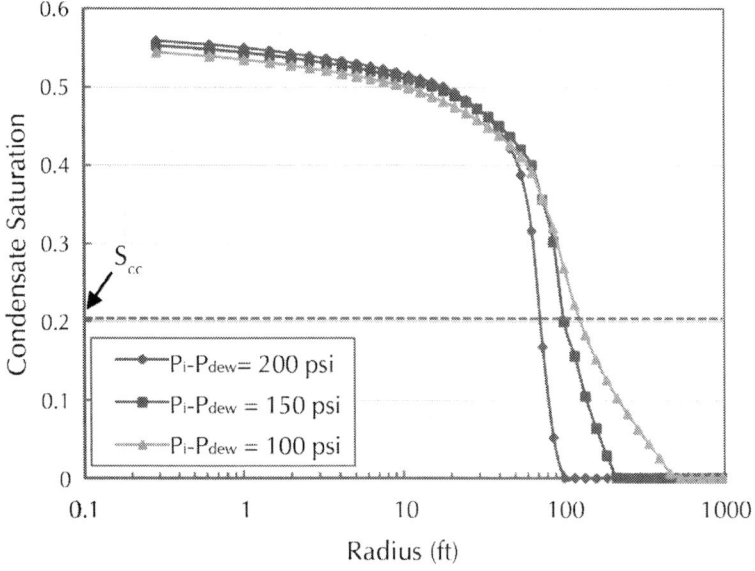

Figure 11: Condensate dropout profile in the reservoir at shut-in time, initial pressure difference effect.

Effect of Production Rate

Four wet gas rates are considered: 8000, 10,000, 12,000 and 15,000 Mscf/day. Condensate dropout profiles are presented in Fig. 13.

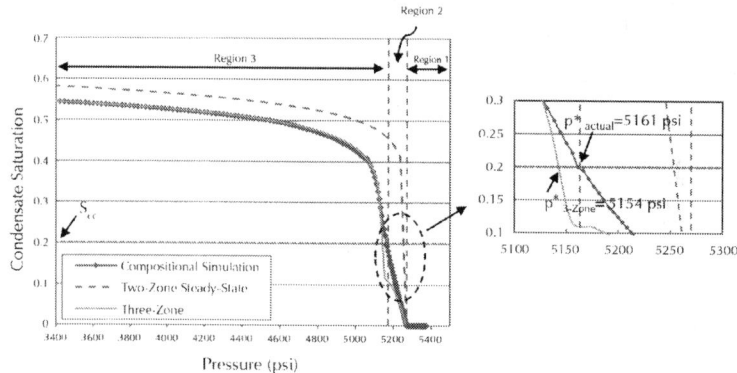

Figure 12: Comparison of condensate saturation from the simulation and as predicted by three-zone and steady-state methods at shut-in time for Mix3 at $p_i - p_{dew} = 100$ psi.

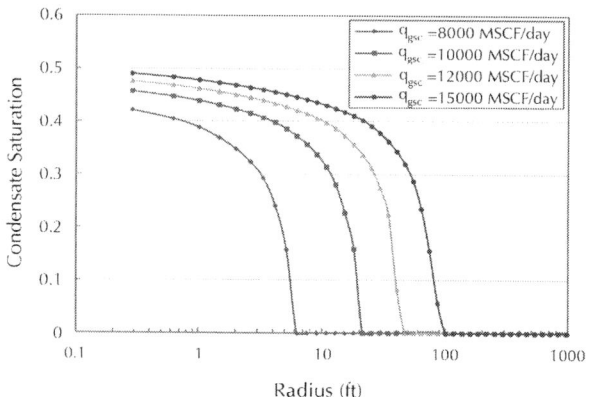

Figure 13: Condensate dropout profile in the reservoir at shut-in time, production rate effect.

The three methods for computing the pseudo-pressure (single-phase, steady-state and three-zone) are tested on a large variety of low permeability cases considering, fluids of different richness in condensate, different relative permeability curves, nonzero skin, different reservoir initial conditions and different flow rate. The effect of each parameter in well testing analysis is discussed in detail and the accuracy of each method in estimating permeability and skin is presented.

Effect of Fluid and Rock Parameters

Fig. 3 shows a comparison of typical single-phase and two-phase pseudo-pressure derivatives plots. According to this figure, in Region 3 where both gas and condensate phases are mobile, after using two-phase pseudo-pressure there is just one stabilization level in the derivative curve, with a small difference between steady-state and three-zone curves. The skin factor and permeability of the reservoir can be calculated from the stabilization levels in this region. Also due to considering only the gas phase in single-phase approach, the corresponding pseudo-pressure derivative curve, lies at a higher level. However single-phase method can be used to estimate the effective gas permeability in this region, hence it may be a useful tool in the well testing analysis of GCRs beside two-phase methods. According to Fig. 3, in Region 2 single-phase and two-phase derivative curve resulted as transition and hump curves, respectively; Region 2 is ignored in two-zone steady-state method and the length of this region is underestimated. Since in this region pseudo-pressure curves are not stabilized, predicted permeability and skin change significantly. Accordingly permeability and skin cannot be predicted from the pseudo-pressure curves in Region 2. Region 1 in which there is a single-phase dry gas fluid flow illustrates the second stabilization curve. This curve which is identical for all methods can be used to calculate skin factor and permeability in this zone. Investigation of different condensate saturation curves with respect to radius from the wellbore (generated using numerical simulator) will indicate the robustness of different parameters in development of condensate bank and condensate dropout profile shape. In the next sections the effect of fluid and rock parameters in formation of condensate bank and the role of these parameters in well test analysis of GCRs are presented.

Fluid Richness

At pressures below the dew-point pressure and at certain conditions of temperature, retrograde condensation occurs in the single-phase fluid and the fluid system separates into two phases: gas phase and liquid phase. Kamath (2007) stated that, a gas condensate system typically yields from about 30 bbl of condensate per MMscf of gas for lean gas condensate and 300 bbl of condensate per MMscf of gas for rich gas condensate. Fig. 4 represents the condensate dropout verses distance for different 4 studied mixtures. According to the figure, fluid richness is a key parameter in the amount of condensate dropout and the radius of condensate bank around the well which causes a decrease in the well productivity.

Fig. 5 shows the condensate saturation profile for the mixture 3 at the time of shut-in. As it is illustrated in this figure, two-phase steady-state method compared to the simulation values overpredicts the condensate saturation whereas three-zone method under predict these values. The circled section in the figure, shows that p^* is underpredicted in three-zone method which results in a lower estimation of the length of Region 3 along with a higher prediction of the length of Region 2. However due to higher estimation of p^* in two-phase steady state method, it tends to underestimate the length of this region. These behaviors with a few differences were observed in all studied cases that may lead to a firm conclusion.

Relative Permeability Curve Inventory

Fig. 6 shows the plot of variation of condensate dropout profile in the reservoir for different cases of relative permeability parameters. In general, three condensate dropout profiles can be observed in this figure:

Case 1

S_{wc}, although decreases the condensate saturation (Fig. 6a), but it has no effect on the volume of condensate formation from the reservoir fluid (Fig. 6b). Considering the effect of connate water saturation, apparently distinct condensate saturation profiles will be observed; a higher amount of connate water yields lower the condensate saturation. Theoretically, if we correct condensate saturation with respect to connate water saturations by using Eq. (9), a similar condensate saturation profile would be achieved in all cases.

Case 2

As it is shown in Fig. 6c–e, S_{cc}, n_g, n_c and K_{rc}^{max} change the value of condensate dropout around a well whereas the radius of condensate bank remains constant. Accordingly well deliverability decreases as a result of higher condensate saturation in the condensate bank region.

Case 3

Fig. 6f shows that end point relative permeability for the gas phase, like richness of fluid as shown in Fig. 4, can change both the value and the radius of condensate dropout. As illustrated in Fig. 6f that is for $S_{cc} = 0.2$, an increase in K_{rg}^{max} decreases the length of Region 2 and the radius of condensate bank around the well with a minor increase in the amount of condensate. In such a situation, the well deliverability can be affected by either the amount or the radius of condensate dropout.

Fig. 7 shows the single-phase and two-phase three-zone pseudo-pressure derivative curves obtained for different three discussed cases. In each case study, other relative permeability parameters are the base case parameters as documented in Table 4.

As it is illustrated, S_{wc} has no impact on pseudo-pressure derivative curve plots. Fig. 7c and d represent the effect of S_{cc} and show that the amount of condensate around the wellbore is increased and it results in an upward shift for the single-phase method in Regions 2 and 3. However, in two-phase method, an addition in the amount of condensate bank increases the height of hump in Region 2 and shifts it forward. Fig. 7e and f indicate the conclusion that can be made from gas end point relative permeability and

shows that a decrease in K_{rg}^{max} affects the radius of condensate bank and as a result, in single-phase method, the length of Region 3 is considerably increased in pseudo-pressure derivative plot. This impact in two-phase pseudo-pressure derivative appears as a forward shift of hump and a reduction in its height due to decrease in the maximum amount of condensate bank around the wellbore.

Mechanical Skin

Fig. 8 shows that the effect of mechanical skin on damaged zone is apparent either in pressure profile or saturation plots. Furthermore, the effect of mechanical skin which is modeled by an alteration in the permeability around the well can be seen in the damaged zone.

As it is illustrated in Fig. 9 in the case of $S = 1.5$, the left hand side of derivative plot depicts a different stabilization line, no matter whether the single-phase or two-phase method has been used. In the other words, in pseudo-pressure derivative plots, the effect of skin can only be seen in the damaged zone. Note that in pseudo-pressure derivative plots, the single-phase method shows effective permeability effect, while two-phase methods reflect absolute permeability effect. As Fig. 10 shows for $S = 3$ which has a larger damaged zone, mechanical skin zone and Region 3 overlap and prediction of skin effect is controversial. In addition, for the negative skin of −1 which may be due to stimulation, the stabilization line is not developed and the distinction of negative skin from pseudo-pressure derivative plots is a difficult task.

Initial Pressure Difference

At the time of discovery, GCRs are often above the dew-point pressure. Accordingly, it's necessary to study the effect of initial reservoir pressure. As it is obvious in Fig. 11, the difference between the initial pressure and dew point pressure ($p_i - p_{dew}$) is an important factor that affects the size of Region 2 and the radius of condensate bank around the well. At greater value of $p_i - p_{dew}$, most of the reservoir life is in a single-phase production period and the reservoir will undergo less pressure drop. In the other words, due to smaller pressure drop, a smaller condensate bank is formed around the wellbore and Region 2 becomes negligible. This phenomena is due to the higher GOR value at a lower $p_i - p_{dew}$, which causes p^* to approach to the dew point pressure (p_{dew}), and consequently liquid condensates become more easily mobile.

Another conclusion that can be made from our observations and discussions of the results obtained in this study on the effect of $p_i - p_{dew}$ on condensate saturation can be drawn once Fig. 12 is compared to Fig. 5, where with only a 100 psi decrease in the value of $p_i - p_{dew}$, the prediction of three-zone method increases significantly ($p^*_{actual} - p^*_{3-zone}$ are 7 and 33 psi, for $p_i - p_{dew}$ of 100 and 200 psi, respectively). The reason backs to the lower $p_i - p_{dew}$ and increases in the condensate bank and the size of Region 2.

Effect of Production Rate

In order to investigate the effect of wet gas production rate, the model was simulated with four rates of 8000, 10,000, 12,000 and 15,000 Mscf/day. Note that production rate is a parameter that can be adjusted in a straightforward manner for the production wells. Fig. 13 illustrates the effect of production rate and shows that once the rate is increased, larger pressure drop is expected, and as a result, the value and the radius of condensate bank is considerably increased and consequently deliverability is reduced. Accordingly the lower production rate is better to control the well deliverability, but it reduces total volume of production. Hence an appropriate production rate should be selected to ensure an optimal recovery from gas condensate wells.

Fig. 14 shows single-phase and two-phase pseudo-pressure derivatives with the equivalent time, obtained for different production rates. As it is apparent, an increase in the production rate changes the height of derivatives plots and shifts them to the right/left direction.

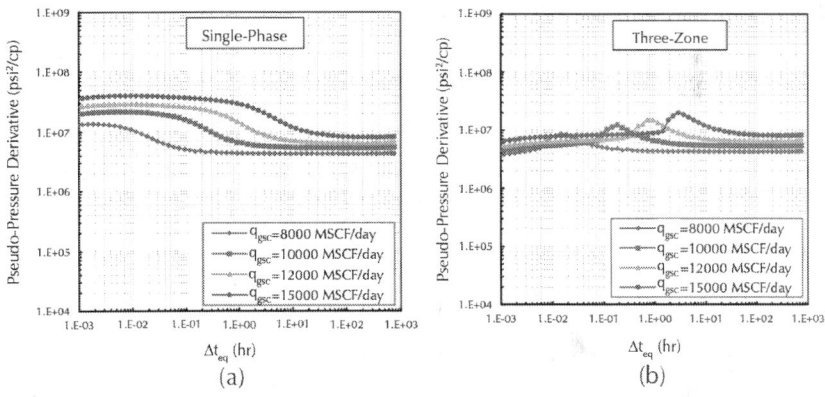

Figure 14: Buildup analysis, (a) single-phase and (b) two-phase three-zone pseudo-pressure derivative, production rate effect.

The Accuracy of Pseudo-pressure Methods

A comprehensive understanding about capability of pseudo-pressure methods in estimating permeability and mechanical skin can be achieved after these parameters were plotted as a function of number of runs. The absolute permeability can be calculated either from single-phase pseudo-pressure in Region 1 or two-phase pseudo-pressure in Region 3. In addition, mechanical skin can be estimated from single-phase pseudo-pressure in Region 3 or two-phase pseudo-pressure in Region 1, while single-phase pseudo-pressure in Region 1 estimated the summation of mechanical and liquid dropout skins. Furthermore, two-phase pseudo-pressure in

Region 3 underestimated mechanical skin by a factor of −1 to −2 for both two-zone and three-zone methods. Note that Region 2 was a transition or hump shaped region in single-phase and two-phase pseudo-pressure derivative curves, respectively and hence it was not stabilized and neither permeability nor skin can be calculated from pseudo-pressure derivative plots in this region.

Fig. 15 shows the ratio of absolute permeability predicted by two-phase methods in Region 3 and single-phase method in Region 1 to the actual value of absolute permeability ($k*/k$) as plotted on the histograms versus number of model run for all 35 cases (the asterisk * refers to the estimated value, no asterisk refers to the defined value in the model). The mean value and the range of $k*/k$ indicated that all three pseudo-pressure methods are capable of fairly accurate permeability estimation (the correct estimation is $k*/k = 1$). However, in some cases, permeability estimates by the two-zone steady-state approach tend to be overestimated (a maximum value of 1.1 for $k*/k$ was obtained). In two-zone steady-state approach, a more accurate mean value of 0.96 for $k*/k$ was evaluated ($k*/k$ values in the single-phase and three-zone methods were obtained 0.9 and 0.88, respectively). Hence the interpretation of simulation cases illustrates that two-zone steady-state method predicts permeability more accurately.

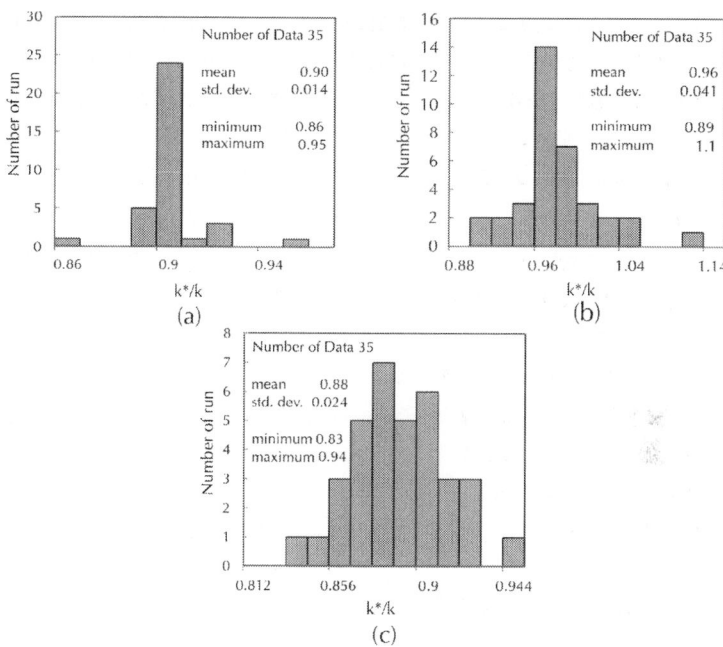

Figure 15: Deviation of permeability from the actual value (*k*/k*) as predicted by (a) single-phase method in Region 1, (b) two-zone and (c) three-zone methods in Region 3 for all studied cases in the buildup analysis.

The results of deviation of estimated mechanical skin from the actual value (*S* − S*) as predicted by single-phase method in Region 3 and two-phase methods in Region 1 for all 35 cases are shown on the histograms in Fig. 16. The least deviation in skin estimates (the correct estimation is *S* − S* = 0) is predicted by the three-zone pseudo-pressure method (a mean value of 0.032 for *S* − S* is obtained for three-zone methods, while single-phase and two-phase steady-state approaches show the values of 0.15 and −0.74, respectively). Furthermore, the skin factor is slightly overestimated by single-phase method (*S* − S* = 0.15), whereas two-zone method significantly underestimates skin (*S* − S* = −0.74). As previously discussed, the skin factor cannot be accurately predicted by two-zone method in Regions 2 and 3, and it is significantly underestimated by this method in Region 3, consequently it cannot be estimated by two-zone steady-state method. Accordingly among the pseudo-pressure methods, three-zone approach can be used to accurately predict the mechanical skin factor in Region 3. Also it can be predicted by single-phase method in Region 1 with a slight overestimation.

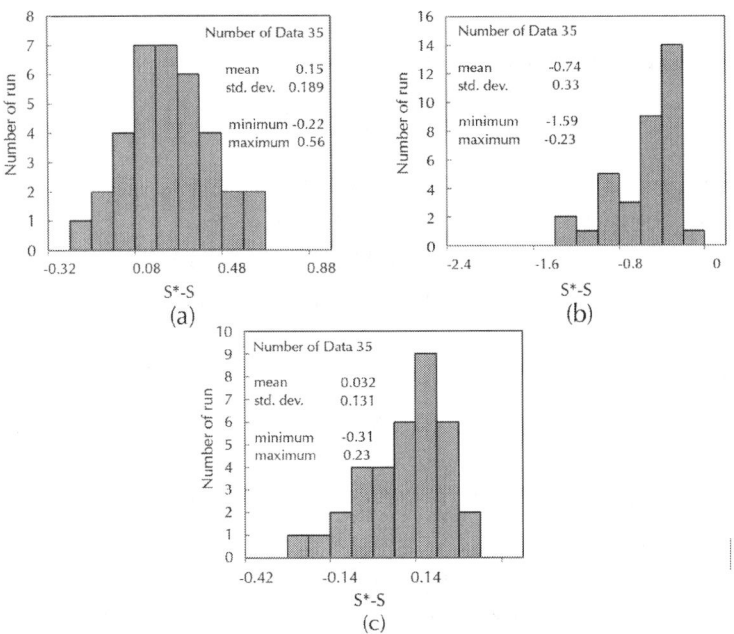

Figure 16: Deviation of skin mechanical factor from the actual value ($S^* - S$) as predicted by (a) single-phase method in Region 3, (b) two-zone and (c) three-zone methods in Region 1 for all studied cases in the buildup analysis.

CONCLUSIONS

In this work, well test and sensitivity analysis were done on 35 different models of low permeability GCRs with the assumption of low capillary number and the following findings are reported:

- Fluid richness is an important factor in determining the amount and radius of the condensate bank around wellbore.

Accordingly lean gas injection can significantly increase well deliverability and consequently improves the recovery of GCRs. And the well test behavior of GCRs is heavily dependent on fluid richness.

- The mechanical skin can affect the well test behavior only in damaged zones and leads to the presence of radial composite behavior in derivative curves of both single-phase and two-phase pseudo-pressure. In the case of a damaged zone with a longer radius, skin zone, Regions 3 and 4 overlap each other and it makes the skin detection difficult.

- The Production rate plays a considerable role in the amount of produced condensate and increasing the radius of condensate bank. Hence, the selection of an appropriate rate controls gas well deliverability loss caused by condensate banking and provides an optimal recovery from gas condensate wells.

- The results of the sensitivity study on relative permeability parameters indicate that:

n_g and n_c the exponents of relative permeability curves that show the fluids miscibility are other important parameters that influence the condensate saturation around wellbore; however, they do not have any effect on the extension of condensate bank radius.

Other parameters, S_{cc} and K_{rc}^{max}, indicate similar effects like n_g and n_c; however, their effects are less. K_{rg}^{max} increases the radius of condensate bank with a little decrease in its saturation. In addition S_{wc} does not have any effect on the separation of condensate from reservoir fluid.

- In two-zone steady-state method, because of disregarding of Region 2, the length of this zone is predicted less than its real length. Thus, p^* comes out more than its real value. This behavior is different in three-zone method and it overestimates the length of Region 2 and calculates p^* less than its actual value.

- Applying the three-zone method for reservoirs with a pressure just a little above their dew point leads to more precise results.

- Results of the single-phase pseudo-pressure indicate that permeability can be correctly estimated in Region 1; and the summation of mechanical and condensate dropout skin factors is generally calculated in this zone. Since the fluid is single phase in Region 1, the results of two single and two-phase pseudo-pressure methods are completely similar. Also, the single-phase method calculates effective permeability in Region 3 and overestimates mechanical skin in this region.

- Since the pseudo-pressure curve in Region 2 is transient for single-phase and hump shaped for two-phase, we cannot use data from this region because of non-stabilization in pseudo-pressure curve. This is one of the drawbacks of pseudo-pressure methods.

- Using the two-zone method is recommended just for permeability calculation in Region 3; however, the three-zone method can calculate skin as well as permeability in this zone.

RECOMMENDATION

According to the conclusions of this research, below recommendations are necessary to be considered:

- Investigating the skin in the initial condition of the reservoir when no condensate is formed is recommended.

- The three-zone method is recommended to calculate skin as well as permeability in Region 3.

REFERENCES

1.   Al-Hussainy, R., Ramey, H., 1966. Application of real gas flow theory to well testing and deliverability forecasting. J. Pet. Tech. 18 (5), 637e642.

2.   Bemani, A.S., Boukadi, F.H., Hajri, R., 2003. Gas-condensate

reservoir pseudorelative permeability derived by pressure transient analysis. Incorporation of near-wellbore effects. J. Pet. Sci. Tech. 21, 1667e1675.

3. Bonyadi, M., Rahimpour, M.R., Esmaeilzadeh, F., 2012. A new fast technique for calculation of gas condensate well productivity by using pseudopressure method. J. Nat. Gas Sci. Eng. 4, 35e43.

4. Fevang, O., Whitson, C.H., 1995. Modeling Gas Condensate Well Deliverability. SPE 30714, Ann. Tech. Conf., Dallas.

5. Fussell, D.D., 1973. Single-well performance predictions for gas-condensate reservoirs. J. Pet. Tech., 860e870.

6. Gringarten, A.C., Bozorgzadeh, M., Daungkaew, S., Hashemi, A., 2006. Well Test Analysis in Lean Gas Condensate Reservoirs: Theory and Practice. SPE 100993 presented at the SPE Russian Oil and Gas Technical Conference and Exhibition.Moscow. Russia.

7. Jamiolahmady, M., Danesh, A., Henderson, G., Tehrani, G.D., 2003. Variation of Gascondensate Relative Permeability with Production Rate at Near Wellbore Conditions: A General Correlation. SPE 83960, Offshore Europe, Aberdeen, UK.

8. Jokhio, S.A., 2002. Production Performance of Horizontal Wells in Gas Condensate

9. Reservoirs. Ph.D. dissertation. University of Oklahoma.

10. Jones, J.R., Vo, D.T., Raghavan, R., 1989. Interoperation of Pressure Buildup Responses in Gas Condensate Wells. SPE Paper No. 15535.

11. Kamath, J., 2007. Deliverability of gas-condensate reservoirs e field experiences and prediction techniques. J. Pet. Tech. 59 (4), 94e99.

12. Liu, J.S., Wilkins, J.R., Al-Qahtani, M.Y., Al-Awami, A.A., 2001. Modeling a Rich Gas Condensate Reservoir with Composition Grading and Faults. SPE 68178 Presented at the 2001 SPE Middle East Oil Show. Bahrain.

13. Mazloom, J., Rashidi, F., 2006. Use of two-phase pseudo

pressure method to calculate condensate bank size and well deliverability in gas condensate reservoirs. J. Pet. Sci. Tech. 24 (2), 145e156.

14.   O'Dell, H.G., Miller, R.N., 1967. Successfully cycling a low-permeability high-yield gas condensate reservoirs. J. Pet. Tech., 41e47.

15.   Xu, S., Lee, J.W., 1999a. Gas Condensate Well Test Analysis Using a Single Phase Analogy. SPE Paper No. 55992.

16.   Xu, S., Lee, J.W., 1999b. Two-phase Well Test Analysis of Gas Condensate Reservoirs. SPE Paper No. 56483.

Analyzing Axial Stress and Deformation of Tubular for Steam Injection Process in Deviated Wells Based on the Varied (T,P) Fields

Yunqiang Liu[1,2], Jiuping Xu[1], Shize Wang[3], and Bin Qi[3]

[1]Uncertainty Decision-Making Laboratory, Sichuan University, Chengdu 610064, China

[2]College of Economics & Management, Sichuan Agricultural University, Chengdu 611130, China

[3]Research School of Engineering Technology, The Southwest Petroleum and Gas Corp, China Petroleum and Chemical Corp, Deyang 618000, China

ABSTRACT

The axial stress and deformation of high temperature high pressure deviated gas wells are studied. A new model is multiple nonlinear equation systems by comprehensive consideration of axial load of tubular string, internal and external fluid pressure, normal pressure between the tubular and well wall, and friction and viscous friction of fluid flowing. The varied temperature and pressure fields were researched by the coupled differential equations concerning mass, momentum, and energy equations instead of traditional methods. The axial load, the normal pressure, the friction, and four deformation lengths of tubular string are gotten by means of the dimensionless iterative interpolation algorithm. The basic data of the X Well, 1300 meters deep, are used for case history calculations. The results and some useful conclusions can provide technical reliability in the process of designing well testing in oil or gas wells.

INTRODUCTION

The deviated wells had been wildly applicable for petroleum and natural gas industry. Deviated wells have their distinctive characteristics which are distinguished from that of other wells. (1) High temperature high pressure: the temperature distribution and pressure on the tubing are significantly different when outputs are varied (flow velocity) but neither has a simple linear relationship, because the fluid density is not constant. (2) Deep well: the sensibility of force and deformation influencing by the factors, such as the temperature, pressure, density of fluid, viscous friction and fluid velocity, and so forth, will become high with the increase of tubing length. The completion test of a deep well is a new problem. In the research of applied basic theory for deep well testing, tubular string mechanical analysis is very complex, but fluid temperature and tubing pressure affect the force of the tubular string heavily. Temperature, pressure, liquid density, and fluid velocity within tubing may change with of the whole depth, time, and operations, so that the axial force changes constantly. A large compression load at low end can induce the tubing plastic deformation and make the packer damaged. A large tension load at the top end may unpack the packer or cause the tubing to break. If the tubing failed, the whole borehole can hardly maintain its integrity and safety [1]. Therefore, it is very important for deviated wells to predict the axial forces for the safety.

Hammerlindl [2] had made a great contribution about tubular mechanics. He had put forth the four effects between the packer forces and length change of tubing: temperature effect, ballooning effect, axial load effect, and the helical buckling effect. There is a large amount of papers to research the effect of buckling behavior. Therefore it is considered that inflexion is caused on its axial force under certain conditions, by which colliding on parts of the drill string with well bore is induced. When buckled of tubular beyond wellhole's control, the buckling configuration which will be transformed at the state of stabilization, sinusoidal buckling and helical buckling with the increase of load. The problem of buckling of the tube was first studied and put into practice by Lubinski et al. [3]. They had done the emulation experiment for the buckling behavior of tube in deviated wells and found the compute formula on critical buckling load of tube in deviated wells. Paslay and Bogy [4] found that the number of sinusoids in the buckling mode increases with the length of the tube. The buckling behavior by inner and outer fluid pressure of tubing was analyzed, and the mathematical relation between pitch and axial pressures was deduced based on the principle of minimum potential energy (see Hammerlindl [2]). The mptotic solution for sinusoidal buckling of an extremely long tube has been analyzed by Dawson and Paslay [5], based on a sinusoidal buckling mode of constant amplitude. Numerical solutions were also sought by Mitchell [6] using the basic mechanics equations. His solutions confirm the thought that, under a general loading, the deformed shape of the tube is a combination of helices and sinusoids while helical deformation occurs only under special values of the applied load. The formula about tubing forces had been put, however, which is too simple for shallow wells to accommodate the complicated states of deep wells. Up to now, many researches are centered on water injection tubular but not on steam injection. Among them, the values of temperature and pressure are considered as constant or lineal functions which will cause large errors on tubular deformation computing [7].

In fact, the tubular string deformation includes transverse deformation and longitudinal deformation. Because the transverse length (its order of magnitude is 10^{-3}m) is much and much smaller than the longitudinal length (its order of magnitude is 10^3m), we mainly consider the axial (longitudinal) deformation for the tubular string deformation analysis in the paper. In the paper, the force states of tubular in the process of steam injection are analyzed. The varied (T, P) fields are considered to compute the values of several deformations. The axial load and four deformation lengths of tubular string are obtained by the dimensionless iterative interpolation algorithm. The basic data of the X Well (deviated well), 1300 meters deep in China, are used for case history calculations. Some useful suggestions are drawn. This paper is organized as follows. Section 2 gives a system model about tubular mechanics and deformation. And the varied (T, P) fields were presented by model concerning mass, momentum, and energy balance. Section 3 gives the parameters, initial condition, and algorithm for solving model. In Section 4, we give an example from a deviated well at 1300 meters of depth in China, and the result analysis are made. Section 5 gives a conclusion.

MODEL BUILDING

Basic Assumption

Before analyzing the force on the microelement, some assumptions are introduced as follows:

- the curvature of the hole of the considered modular section is constant,
- on the upper side or underside of the section which is point

of contact of the pipe and tube wall, the curvature is the same with the hole curvature,

- the radius of steam injection string, in contrast to curvature of borehole, is insignificant,
- the string is at the state of linear elastic relationship.

Forces Analysis of Tubular String

The forces of tubular string are shown in Figure 1. Consider the flow system depicted in Figure 1: a constant cross-sectional flow area A, inner diameter d, outer diameter D, material density ρ_1, packer fluid density ρ_2, and a total length Z. Through this tubing gas flows from the bottomto the topwith a mass flow rate W. The distance coordinate in the flow direction along the tubing is denoted by z.The cylindrical coordinate system $r\theta z$, origin of which is in wellhead and Z axis is down as the borehole axis, is used.

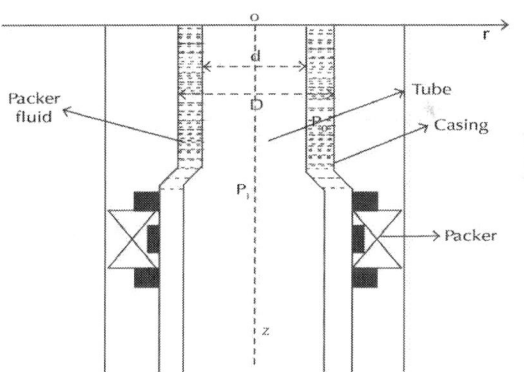

Figure 1: The physical figure of forces analysis on tube.

As shown from Figure 1, the tubular string is mainly acted upon by the following forces at the process of steam injection.

- Initial Axial Force. The initial axial force of tubular should include the deadweight, buoyant weight, and initial pull force.

- Thermal Stress. On the process of steam injection, the temperature stress will act at the tubular with varied temperature.
- Axial Force by the Varied Internal and External Pressure. Thanks to the varied pressure with internal and external pressure, the tubular will be acted by the bending force, piston force, and other axial forces.
- Friction Drag by Steam Injection. On the process of steam injection, the flow in tubular will produce viscous flow which will cause the friction drag.

The Axial Load and Axial Stress of the Tubular

Initial Axial Load and Initial Axial Stress of Steam Injection Tubular

Initial Axial Load. The section to which the distance from the wellhead is z (m) was considered. The axial static load by the deadweight of tubular is as follows:

$$N_{qz} = \int_z^L q\cos\alpha\,dz = \frac{\pi}{4}\rho_1 g\left(D^2 - d^2\right)\int_z^L \cos\alpha\,dz, \tag{1}$$

where N_{bz} is the deadweight of tubular, q is the average unit length weight of tubing, L is the length of tubular, ρ_1 is the density of tubular, and α is the inclination angle.

The axial static load by the buoyant weight is as follows:

$$N_{bz} = -\rho_2 g A_2 \int_z^L \cos\alpha\,dz = -\rho_2 g z \pi\left(\frac{D}{2}\right)^2 \int_z^L \cos\alpha\,dz, \tag{2}$$

where N_{pz} is the buoyant weight of tubular, ρ_2 is the density of packer fluid.

The axial load by the steam injection pressure

$$N_{pz} = \frac{P_{zI} \pi d^2 z}{4},$$

(3)

where, P_{zI} represents the inner pressure at this section.

Therefore, summing (1), (2), and (3), the axial forces in the section are obtained as follows:

$$F_z = N_{qz} + N_{bz} + N_{pz}.$$

(4)

Initial Axial Stress. The axial stress can be derived from the following equation

$$\sigma_{zi} = \frac{4F_z}{\pi (D^2 - d^2)}.$$

(5)

Axial Thermal Stress of Steam Injection Tubular

In the process of steam injection, the temperature of tubular will change with time and depth, which will make the tubular deform as follows:

$$\sigma_{zt} = E\beta (T_{z1} - T_{z0}) = E\beta \Delta T,$$

(6)

where, E represents the steel elastic modulus of tubular, β is the warm balloon coefficient of the tubular string, and ΔT is the temperature change with before and after steam injection.

Axial Stress of Steam Injection Tubular by the Change with Pressure

The effect acting the tubular with pressure change which is called ballooning effect normally.

Ballooning Stress Analysis. The ballooning effect will be produced from pressure acted in inner and outer of the tube. Generally, there are two kinds of tubular in oil wells. One is the tubulars whose outer diameter is 88.9 mm, inner diameter is 76 mm, and thickness of tubes is 6.5 mm ($\delta/(d/2) = 17.1\% > 5\%$) the other is the tubular whose outer diameter is 114.3 mm, inner diameter is 100.5 mm, and thickness of tubes is 6.9 mm ($\delta/(d/2) = 13.7\% > 5\%$). Neither is the thin-wall problem. Therefore, it should be solved by Lame's formula [8].

The radial and tangential stresses in the thick-wall cylinder can be shown as Figure 2. The two can be calculated as follows:

$$\sigma_{rz} = \frac{d^2 P_{z1} - D^2 P_{z0}}{D^2 - d^2} - \frac{(P_{z1} - P_{z0}) D^2 d^2}{(D^2 - d^2) 4r^2},$$

$$\sigma_{\theta z} = \frac{d^2 P_{z1} - D^2 P_{z0}}{D^2 - d^2} + \frac{(P_{z1} - P_{z0}) D^2 d^2}{(D^2 - d^2) 4r^2},$$

$$(7)$$

where r is radial stress, θ is tangential stress, r $(d \leq r \leq D)$ is radial coordinate, P_{z1} is tube internal pressure at z point, and P_{z0} is tube external pressure at z point.

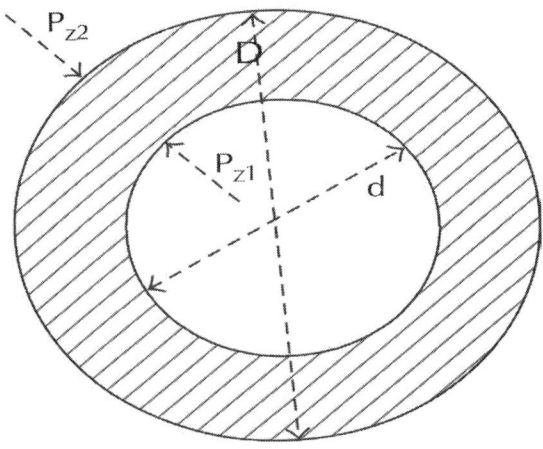

Figure 2: The radial and tangential stresses figure of tube.

Axial Stress of Steam Injection Tubular by the Friction Loss

In fact, the flow in the tubular should be multiflow. On the process of steam injection, the flow will be run and it will give rise to friction effect to cause axial stress. In our paper, we consider the flow gas-liquid mix flow and the liquid head loss is gotten by the Darcy-Weisbach formula [9] as follows:

$$h_f = \frac{\lambda (Z - z) v_m^2}{2gd},$$

(8)

where h_f means heat loss of liquid flow, λ is frictional head losses coefficients, and V_m is the velocity of liquid flow.

The friction drag in tubular is $N_{fz} = h_f \rho_m g \pi d^2$ (ρ_m is density of liquid flow).The axial stress by fiction drag can be obtained as follows:

$$\sigma_{zf} = \frac{4N_{fz}}{\pi (D^2 - d^2)}.$$

(9)

Analysis of Axial Deformation

Based on the studies and analyses mentioned above, the axial deformation on the tubular is made up of the following parts.

The Axial Deformation by the Axial Static Stress

For the microelement of the tubular dz, the unit deformation by the static stress can be computed by generalized Hooke law

$$\varepsilon_1 = \frac{1}{E} \left[\sigma_{zi} - \mu \left(\sigma_{rz} + \sigma_{\theta z} \right) \right],$$

(10)

where μ represents Poisson's ratios.

The axial deformation at an element can be obtained through integrating on the length of the element as follows:

$$\Delta L_{1i} = \int_{Z_{i-1}}^{Z_i} \frac{1}{E} \left[\sigma_{zi} - \mu \left(\sigma_{rz} + \sigma_{\theta z} \right) \right] dz.$$

(11)

Therefore, the total axial deformation by the static stress can be gotten accumulating each element as follows:

$$\Delta L_1 = \sum_{i=1}^{N} \Delta L_{1i}.$$

(12)

The Axial Deformation with Temperature Changed

For the microelement of the tubular dz, the unit deformation by the temperature change is as follows:

$$\Delta L_{2i} = \int_{Z_{i-1}}^{Z_i} \frac{\sigma_{zt}}{E} dz = \beta \Delta T_i \Delta L_i.$$

(13)

The same principle is that the total axial deformation by the varied temperature fields can be gotten accumulating each element as follows:

$$\Delta L_2 = \sum_{i=1}^{N} \Delta L_{2i}.$$

(14)

The Axial Deformation with the Friction Drag

For the microelement of the tubular dz, the unit deformation by the friction force is as follows:

$$\Delta L_3 = \int_0^Z \frac{\sigma_{zf}}{E} dz = \frac{\lambda \rho_m v_m^2 dZ^2}{E(D^2 - d^2)}.$$

(15)

The Axial Deformation with the Tubular String Buckling

Researchers in general call the buckling a bending effect. The tubular is freely suspended in the absence of fluid inside as shown in Figure 3(a). Because the force F applied at the end of the tubular which is large enough, the tubular will buckle as shown in Figure 3(b).

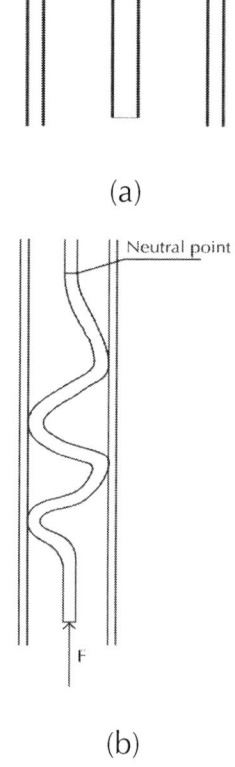

(a)

(b)

Figure 3: Buckling of tubular.

Lubinski et al. [3] had done many researches on the phenomenon. From their work, we can get the buckling effect. Define the virtual axial force of tubing as follows:

$$F_f = A_p (P_1 - P_0),$$

(16)

where P_1 is the pressure inside the tubular at the packer length, P_0 is the pressure outside the tubular at the packer length, and A_p is the area corresponding to packer bore.

By (16), whether the tubular will buckle or not can be judged. The string will buckle if F_f is positive or remain straight if F_f is negative or zero. The axial deformation of the tubular string buckling is

$$\Delta L_{4i} = -\frac{r^2 A_p^2 (\Delta P_{1i} - \Delta P_{0i})^2}{8EIW},$$

(17)

where r means tubing-to-casing radial clearance, I is moment of inertia of tubing cross-section with respect to its diameter ($I = \pi(D^4 - d^4)/64$), Δ denotes change with before and after injection, and W is the unit weight of tubing, as

$$\Delta L_4 = \sum_{i=1}^{N} \Delta L_{4i}.$$

(18)

In addition, the position of the neutral point is needed. The length (n) from the packer to the point can be computed as follows:

$$n = \frac{F_f}{W}.$$

(19)

Generally, the neutral point should be in tubular ($n \leq Z$). However, at the multipackers, it will occur that the neutral point is outside the tubing between dual packers. In this paper, we leave the latter phenomenon.

To sum up, the whole deformation length can be represented as follows:

$$\Delta L = \Delta L_1 + \Delta L_2 + \Delta L_3 + \Delta L_4.$$

(20)

The Analysis of the Varied (T,P) Fields

In the course of dryness modeling, we can find that the numerical values of deformation ((10), (13), and (17)) were affected by the temperature and pressure. In fact, the two parameters varied according to the depth and time changing. So, the varied (T, P) fields need to be researched. Under the China Sinopec Group Hi-Tech Project "Stress analysis and optimum design of well completion" in 2009 [6] undertaken by Sichuan University at early time. The varied (T, P) fields had been deduced strictly based on the mass, momentum, and energy balance. The proof details can be shown in Xu et al. [11]. The varied (T, P) fields is

$$\frac{dP}{dz} = \frac{-(\tau_i/A) + \rho_m g \cos\theta + (m/A) R (dx/dz)}{1 - (m/A) S},$$

$$\frac{dT}{dz} = -\frac{v_m}{C_{Pg}} \left(R\frac{dx}{dz} - S\frac{dP}{dz} \right) - \frac{g \cos\theta}{C_{Pg}}$$

$$- \frac{\pi f r_{ti} \rho_m v_m^3}{4C_{Pg}} + \frac{a(T - T_e)}{C_{Pg}},$$

$$P(z_0) = P_0, \quad T(z_0) = T_0, \quad dx(z_0) = dx_0, \quad x(z_0) = z_0.$$

(21)

NUMERICAL IMPLEMENTATION

Calculation of Some Parameters

In this section, we will give the calculating method of some parameters.

- Each point's inclination:

$$\alpha_j = \alpha_{j-1} + \frac{(\alpha_k - \alpha_{k-1}) \Delta s_j}{\Delta s_k},$$

(22)

where j represents segment point of calculation, Δ_{sk} represents measurement depth of inclination angle α_k, and α_{k-1}, Δ_{sj} is the step length of calculation. Transient heat transfer function [12]:

$$f(t_D) = \begin{cases} 1.128\sqrt{t_D}(1 - 0.3\sqrt{t_D}), & t_D \leq 1.5, \\ (0.4063 + 0.5\ln t_D)\left(1 + \frac{0.6}{t_D}\right), & t_D > 1.5. \end{cases}$$

(23)

- The density of wet steam. Since the flow of the water vapor in is the gas-liquid two-phase flow, there are many researches about this problem [13, 14]. In the paper, we adopt the M-B model to calculate the average density of the mixture.

- The heat transfer coefficient U_{to} from different positions of the axis of the wellbore to the second surface.

These resistances include the tubing wall, possible insulation around the tubing, annular space (possibly filled with a gas or liquid but is sometimes vacuum), casing wall, and cementing behind the casing as follows:

$$\frac{1}{U_{to}} = r_{ti}\frac{1}{\lambda_{ins}}\ln\left(\frac{r_{ci}}{r_{to}}\right) + \frac{1}{h_c + h_r} + r_{ti}\frac{1}{\lambda_{cem}}\ln\left(\frac{r_{cem}}{r_{co}}\right)$$

(24)

λ_{ins} and λ_{cem} are the heat conductivity of the heat insulating material and the cement sheath, respectively. h_c and h_r are the coefficients of the convection heat transfer and the radiation heat transfer.

Initial Condition

In order to solve model, some definite conditions and initial conditions should be added. The initial conditions comprise the distribution of the pressure and temperature at the well top. In this paper, we adopt the value at the initial time by actual measurement.

Before steam injected, the temperature of tubular just is initial temperature of formation ($T_z = T_0 + \gamma z \cos \alpha$, γ is geothermal gradient). At the same time, the pressure of inner tubular is assumed to be equal to the outer tubular before steam injected.

Steps of Algorithm

To simplify the calculation, we divided the wells into several short segments of the same length. The length of a segment varies depending on variations in wall thickness, hole diameter, fluid density inside and outside the pipe, and wells geometry. The model begins with the calculation at one particular position in the wells: the top of the pipe.

- Step 1. Set step length of depth. In addition, we denote the relatively tolerant error by ε. The smaller h, ε is, the more accurate the results are. However, it will lead to rapid increasing calculating time. In our paper, we set $h = 1$(m), and $\varepsilon = 5\%$.
- Step 2. Give the initial conditions.
- Step 3. Compute each point's inclination.
- Step 4. Compute the parameters under the initial conditions or the last depth variables.
- Step 5. Let $T = T_k$; then we can get the T_e by solving the following equation:

$$\frac{\partial T_e}{\partial t_D} = \left(\frac{\partial^2 T_e}{\partial r_D^2} + \frac{1}{r_D} \frac{\partial T_e}{\partial r_D} \right),$$

$$T_e\big|_{t_D=0} = T_0 + \gamma z \cos \theta,$$

$$\frac{\partial T_e}{\partial r_D}\bigg|_{r_D=1} = -\frac{1}{2\pi \lambda_f} \frac{dq}{dz},$$

$$\frac{\partial T_e}{\partial r_D}\bigg|_{r_D \to \infty} = 0.$$

(25)

Let T_e^j be the temperature at the injection time j and radial i at the depth z. We apply the finite different method to discretize the equations as follows:

$$\frac{T_{e,j}^{i+1} - T_{e,j}^i}{\varphi} = \frac{T_{e,j+1}^{i+1} - 2T_{e,j+1}^j + T_{e,j+1}^{i-1}}{\xi^2} - \frac{T_{e,j+1}^{i+1} - T_{e,j}^{i+1}}{r_D \varphi},$$

(26)

where φ is the interval of time and ξ is the interval of radial, respectively. It can be transformed into the standard form as follows:

$$-\left(\varphi + \frac{\varphi\xi}{r_D}\right)T_{e,j+1}^{i+1} + \left(2\varphi + \frac{\varphi\xi}{r_D}\right)T_{e,j}^{i+1} - \varphi T_{e,j-1}^{i+1} = \xi^2 T_{e,j}^i.$$

(27)

Then the different method is used to discretize the boundary condition. For $r_D=1$, we have

$$T_{e,2}^{e,i+1} - \left(1 + \frac{a\xi}{2\pi\lambda_f}\right)T_{e,1}^{i+1} = \frac{aT_k}{2\pi\lambda_f}.$$

(28)

For $r_D=N$, we have

$$T_{e,n}^{i+1} - T_{e,n-1}^{i+1} = 0.$$

(29)

We can compute the symbolic solution of the temperature T_e of the stratum. In this step, we will get the discrete distribution of T_e as the following matrix:

$$\begin{bmatrix} T_{e,1}^1 & T_{e,1}^2 & \cdots & T_{e,1}^i & \cdots \\ T_{e,2}^1 & T_{e,2}^2 & \cdots & T_{e,2}^i & \cdots \\ \vdots & \vdots & \cdots & \vdots & \vdots \\ T_{e,j}^1 & T_{e,j}^2 & \cdots & T_{e,j}^i & \cdots \\ \vdots & \vdots & \cdots & \vdots & \vdots \\ T_{e,n}^1 & T_{e,n}^2 & \cdots & T_{e,n}^i & \cdots \end{bmatrix},$$

(30)

where i represents the injection time and j represents the radial.

Step 6. Let the right parts of the coupled differential equations be functions F_k, where (K=1,2). Then we can obtain a system of coupled functions as follows:

$$F_1 = \frac{-(\tau_i/A) + \rho_m g \cos\theta + (m/A) R\,(dx/dz)}{1 - (m/A) S},$$

$$F_2 = -\frac{\nu_m}{C_{Pg}}\left(R\frac{dx}{dz} - S\frac{dP}{dz}\right) - \frac{g\cos\theta}{C_{Pg}}$$

$$-\frac{\pi f r_{ti} \rho_m \nu_m^3}{4C_{Pg}} + \frac{a\,(T - T_e)}{C_{Pg}},$$

$$\tag{31}$$

where T_e at $r_D = 1$.

Step 7. Assume that P,T are y_k (k=1,2), respectively. Then we can obtain some basic parameters as follows:

$$a_k = F_i\,(y_1, y_2),$$

$$b_k = F_i\left(y_1 + \frac{ha_1}{2}, y_2 + \frac{ha_2}{2}\right),$$

$$c_k = F_i\left(y_1 + \frac{hb_1}{2}, y_2 + \frac{hb_2}{2}\right),$$

$$d_k = F_i\,(y_1 + hc_1, y_2 + hc_2).$$

$$\tag{32}$$

Step 8. Calculate the pressure and temperature at point (j+1):

$$y_k^{j+1} = y_k^j + \frac{h\,(a_k + 2b_k + 2c_k + d_k)}{6},$$

$$k = 1, 2, \quad j = 1, 2, \ldots, n.$$

$$\tag{33}$$

Step 9. Calculate the deformation ΔL_{1j}, ΔL_{2j}, and ΔL_{4j} by previous equations.

Step 10. Repeat the third step to the tenth step until tubular length Z is calculated.

Step 11. Calculate the deformation ΔL_3 and total deformation length as follows:

$$\Delta L = \sum_{j=1}^{N} \Delta L_{1j} + \sum_{j=1}^{N} \Delta L_{2j} + \Delta L_3 + \sum_{j=1}^{N} \Delta L_{4j}.$$

(34)

NUMERICAL SIMULATION

Parameters

To demonstrate the application of our theory, we study a pipe in X well, which is in Sichuan Province, China. All the basic parameters are given as follows: depth of the well is 1300 m; ground thermal conductivity parameter is 2.06; ground temperature is 16°C; ground temperature gradient is 0.0218(°C/m); roughness of the inner surface of the well is 0.000015; and parameters of pipes, inclined well, inclination, azimuth, and vertical depth are given in Tables 1, 2, and 3.

Table 1: Parameters of pipes

Diameter(m)	Thickness(m)	Weight(Kg)	Expansion	Elastic(Gpa)	Poisson's ratios	Using length(m)
0.0889	0.01295	23.79	0.0000115	215	0.3	270
0.0889	0.00953	18.28	0.0000115	215	0.3	120
0.0889	0.00734	15.04	0.0000115	215	0.3	620
0.0889	0.00645	13.58	0.0000115	215	0.3	290

Table 2: Well parameters

Measured (m)	Internal (m)	External (m)
336.7	0.15478	0.1778
422.6	0.1525	0.1778
1300.0	0.10862	0.127

Table 3: Parameters of azimuth, inclination, and vertical depth

Number	Measured(m)	Inclination(°)	Azimuth(°)	Vertical depth(m)
1	135	2.63	241.01	134.72
2	278	1.23	237.86	277.91
3	364	1.43	213.86	363.82
4	393	2.17	26.38	392.53
5	422	1.85	44.56	421.28
6	450	0.82	191.12	449.62
7	486	2.93	269.07	485.47
8	514	1.03	297.55	513.83
9	543	3.58	324.51	541.74
10	571	2.98	303.05	570.43
11	600	2.03	204.74	599.42
12	628	2.34	164.33	627.28
13	660	1.85	195.28	659.56
14	723	3.14	214.84	721.70
15	782	0.98	216.48	781.30
16	830	2.15	229.31	829.12
17	860	2.67	244.03	859.71
18	908	4.85	266.62	904.08
19	928	6.72	258.78	921.42
20	972	2.03	236.88	971.71
21	1025	4.78	239.27	1021.25
22	1058	4.01	244.59	1055.58
23	1089	4.98	228.2	1084.17
24	1132	3.75	233.88	1129.28
25	1174	5.63	235.14	1168.87
26	1204	4.23	234.38	1200.99
27	1235	3.87	234.99	1232.08
28	1268	4.97	232.57	1263.45
29	1300	8.84	233.28	1284.96

Main Results and Results Analysis

After calculation, we obtain a series of results of this well as Table 4. The influence of outputs on the axial deformation of tubing was investigated as shown by Figure 4.

Table 4: The results of the axial force and various kinds of deformation lengths

Number	Depth(m)	Axial force(N)	Displacement by temperature changed (m)	Displacement by pressure changed (m)	Axial deformation(m)	Buckling deformation (m)	Total deformation(m)
1	1	895244.8	0	0	0	0	0
2	100	854724.8	0.1201	0.00986	0.024	0	0.1544
3	200	814215.5	0.2362	0.019392	0.052	0	0.3072
4	300	773717.7	0.3483	0.028598	0.082	0	0.459
5	400	737970	0.4564	0.037476	0.115	−0.006	0.6029
6	500	706877.3	0.5606	0.046028	0.152	−0.006	0.7523
7	600	675763.9	0.6607	0.054254	0.192	−0.006	0.9006
8	700	644602.3	0.7569	0.062153	0.235	−0.006	1.048
9	800	613437.2	0.849	0.069725	0.283	−0.007	1.1946
10	900	582272.1	0.9371	0.076968	0.335	−0.007	1.3422
11	1000	551107.2	1.0212	0.083883	0.391	−0.007	1.4896
12	1100	519942.3	1.1014	0.090471	0.452	−0.007	1.6367
13	1200	488777.5	1.1775	0.096731	0.517	−0.009	1.7822
14	1300	457612.8	1.2496	0.102662	0.584	−0.01	1.9261

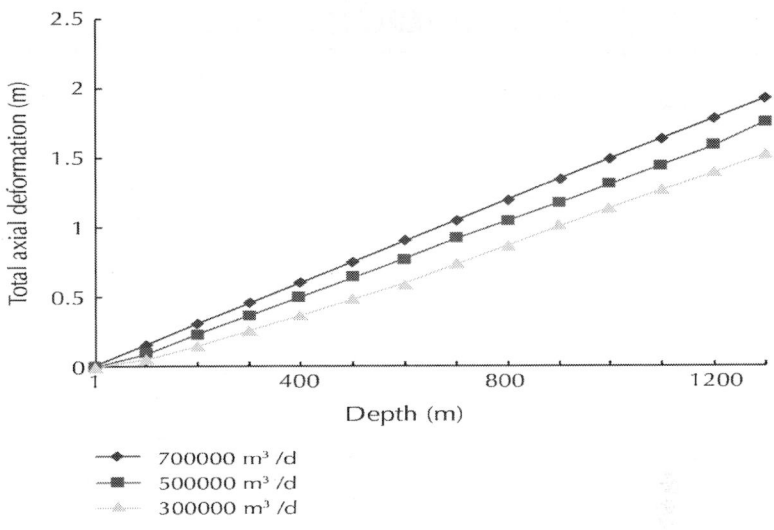

Figure 4: The total axial deformation under varied outputs.

From the results as shown in Figure 4 and Table 4, some useful analysis can be drawn.

- The amount of steam injected and inject pressure affected the stretching force with special severity.
- The results were as follows: the length of tubular deformation was risen with increased injected pressure or injected velocity.
- The length of tubular deformation increases with the increasing of outputs but more slowly.
- The thermal stress is the main factor influencing the tubular deformation. Therefore, the temperature of steam injected should not be too high.
- The lifting prestressed cementing technology has important meanings to reduce the deformation of tubular.
- The creeping displacement of downhole stings will produce an upward contractility which causes packer depressed or lapsed. Therefore, the effective measures should be adopted to control the companding of tubular.

CONCLUSIONS

In this paper, a total tubular deformation model about deviated wells was given. A coupled-system model of differential equations concerning pressure and temperature in high temperature-high pressure steam injection wells according to mass, momentum, and energy balances, which can reduce the error of axial stress and axial deformation, was given instead of the average value or simple linear relationship in traditional research. The basic data of the Well (high temperature and high pressure gas well), 1300 m deep in Sichuan, China, were used for case history calculations. The results can provide technical reliance for the process of designing well tests in deviated gas wells and dynamic analysis of production.

ACKNOWLEDGMENTS

This research was supported by the Key Program of NSFC (Grant no. 70831005) and the Key Project of China Petroleum and Chemical Corporation (Grant no. GJ-73-0706).

REFERENCES

1.   D.-L. Gao and B.-K. Gao, "A method for calculating tubing behavior in HPHT wells," Journal of Petroleum Science and Engineering, vol. 41, no. 1–3, pp. 183–188, 2004

2.   D. J. Hammerlindl, "Movement, forces, and stresses associated with combination tubing strings sealed in packers," Journal of Petroleum Technology, vol. 29, pp. 195–208, 1977

3.   A. Lubinski, W. S. Althouse, and J. L. Logan, "Helical buckling of tubular sealed in packers," Journal of Petroleum Technology, vol. 14, no. 6, pp. 655–670, 1962

4.   P. R. Paslay and D. B. Bogy, "The stability of a circular rod laterally constrained to be in contact with an inclined circular

cylinder," Journal of Applied Mechanics, vol. 31, pp. 605–610, 1964

5.   R. Dawson and P. R. Paslay, "Drillpipe buckling in inclined holes," Journal of Petroleum Technology, vol. 36, no. 10, pp. 1734–1738, 1984

6.   R. F. Mitchell, "Effects of well deviation on helical buckling," SPE Drilling and Completion, vol. 12, no. 1, pp. 63–68, 1997

7.   P. Ding and X. Z. Yan, "Force analysis of high pressure water injection string," Petroleum Dring Techiques, vol. 36, no. 5, p. 23, 2005

8.   Z. F. Li, "Casing cementing with half warm-up for thermal recovery wells," Journal of Petroleum Science and Engineering, vol. 61, no. 2–4, pp. 94–98, 2008

9.   A. M. Sun, "The analysis and computing of water injection tubular," Drilling and Production Technology, vol. 26, no. 3, pp. 55–57, 2003 (Chinese

10.   J. P. Xu, "Stress analysis and optimum design of well completion," Technical Report of Sinopec GJ-73-0706, 2009

11.   J. P. Xu, Y. Q. Liu, S. Z. Wang, and B. Qi, "Numerical modelling of steam quality in deviated wells with variable (T, P) fields," Chemical Engineering Science, vol. 84, pp. 242–254, 2012

12.   A. R. Hasan and C. S. Kabir, "Two-phase flow in vertical and inclined annuli," International Journal of Multiphase Flow, vol. 18, no. 2, pp. 279–293, 1992

13.   H. D. Beggs and J. R. Brill, "A study of two-phase flow in inclined pipes," Journal of Petroleum Technology, vol. 25, no. 5, pp. 607–617, 1973, paper 4007-PA

14.   H. Mukherjee and J. P. Brill, "Pressure drop correlations for inclined two-phase flow," Journal of Energy Resources Technology, vol. 107, no. 4, pp. 549–554, 1985

Citations

CHAPTER 1

Kashif Rashid, William Bailey, and Benoît Couët, "A Survey of Methods for Gas-Lift Optimization," Modelling and Simulation in Engineering, vol. 2012, Article ID 516807, 16 pages, 2012. doi:10.1155/2012/516807.

CHAPTER 2

G.J. Moridis, S. Silpngarmlert, M.T. Reagan, T. Collett, K. Zhang, Gas production from a cold, stratigraphically-bounded gas hydrate deposit at the Mount Elbert Gas Hydrate Stratigraphic Test Well, Alaska North Slope: Implications of uncertainties, Marine and Petroleum Geology, Volume 28, Issue 2, February 2011, Pages 517-534, ISSN 0264-8172, doi.org/10.1016/j.marpetgeo.2010.01.005.

CHAPTER 3

Longjun Zhang, Daolun Li, Lei Wang, and Detang Lu, "Simulation of Gas Transport in Tight/Shale Gas Reservoirs by a Multicomponent Model Based on PEBI Grid," Journal of Chemistry, Article ID 572434, in press.

CHAPTER 4

Brian Anderson, Steve Hancock, Scott Wilson, Christopher Enger, Timothy Collett, Ray Boswell, Robert Hunter, Formation pressure testing at the Mount Elbert Gas Hydrate Stratigraphic Test Well, Alaska North Slope: Operational summary, history matching, and interpretations, Marine and Petroleum Geology, Volume 28, Issue 2, February 2011, Pages 478-492, ISSN 0264-8172, http://dx.doi.org/10.1016/j.marpetgeo.2010.02.012.

CHAPTER 5

Dzeti Farhah Mohshim, Hilmi bin Mukhtar, Zakaria Man, and Rizwan Nasir, "Latest Development on Membrane Fabrication for Natural Gas Purification: A Review," Journal of Engineering, vol. 2013, Article ID 101746, 7 pages, 2013. doi:10.1155/2013/101746.

CHAPTER 6

Hossein Mohammadi, Mohammad Hossein Sedaghat, Abbas Khaksar Manshad, Parametric investigation of well testing analysis in low permeability gas condensate reservoirs, Journal of Natural Gas Science and Engineering, Volume 14, September 2013, Pages 17-28, ISSN 1875-5100, http://dx.doi.org/10.1016/j.jngse.2013.04.003.

CHAPTER 7

Yunqiang Liu, Jiuping Xu, Shize Wang, and Bin Qi, "Analyzing Axial Stress and Deformation of Tubular for Steam Injection Process in Deviated Wells Based on the Varied (T,P) Fields," The Scientific World Journal, vol. 2013, Article ID 565891, 9 pages, 2013. doi:10.1155/2013/565891.

Index

C

Constant composition expansion (CCE)
189

D

Deviated wells 218, 219, 238, 239
Dynamic-programming-(DP-) 27

F

Flowing bottom-hole pressure (FBHP)
136, 137, 140, 141

G

Gas condensate reservoirs (GCRs) 185
Gas-lift injection rate (GLIR) 2
Gas-lift performance curve (GLPC) 7,
10
Gas-lift valve (GLV) 4
Gas liquid ratios (GLR) 10
Gas production 46, 47, 49, 51, 52, 55, 61,
64, 67, 73, 78, 85, 90, 91, 92, 96, 97
Generate well relationships (GLPC) 15
Genetic algorithm (GA) 22
Geochemistry 52
GMRES (generalized minimal residual
method) 108

H

Hydrate-bearing layer (HBL), 49
Hydrates 51, 55, 63, 71, 80, 85, 92, 96

I

Incremental gas-oil-ratio (IGOR) 12
Inflow performance curve (IPR) 10

K

Karush-Kuhn-Tucker (KKT) 17

L

Liquid supported membrane (ILSM) 172
Lohrenz-Bray-Clark (LBC) 106

M

Methane 51, 95
Mixed-integer linear programming
    (MILP) 27
Mixed-integer non-linear (MINLP) 34
Mixed-integer non-linear program
    (MINLP) 27
Mixed matrix membrane (MMM) 164,
    170, 171
Modular Dynamic Tester (MDT) 127
Modular Dynamic Testing (MDT) 52
Multi-objective problem (MOP) 23

N

Neural Net (NN) 25
Newton Reduction Method (NRM) 13,
    18

O

Non-instantaneous flow (NIF) 20, 21
Non-instantaneously flowing (NIF) 18

O

Operating cash income (OCI) 10, 13

P

PEBI (perpendicular bisection) 100, 101
Peng-Robinson equation of state (PR
    EOS) 106
Permafrost-associated 92
Plunger gas-lift (PL) 4
Polyamide (PI) 169
Polyetehrsulfone (PES) 169
Polysulfone (PSU) 168, 169
Present value operating cash income
    (PVOCI) 13
Pressure volume temperature (PVT) 192
Productivity index (PI) 17

R

Room temperature ionic liquid (RTIL)
    172

S

Sequential Linear Programming (SLP)
    18
Sequential Quadratic Programming
    (SQP) 19

W

Wellhead pressures (WHP) 26